Multiplication Is Fun!

(and division is delightful)

Created by
Emily Jacques

©Copyright 2016 by Emily Jacques. All rights reserved. See note to parents and teachers for further clarification.

****

I dedicate this book to all the children and grandchildren of my former students, who will all hopefully have better math teachers than I was to their ancestors. ;)

****

Acknowledgements: Many thanks to my husband, Jerry, for helping his hypersensitive-to-light wife with the formatting and some of the graphics in this book.

Thanks to Kevin McLeod, owner of incompetech.com, where he has generously provided any kind of graph paper you want for free for years.

Thanks to the several math-related websites who provided me with things like a multiplication table.

Thanks to all the teachers of all the teacher workshops I attended over my school-teaching career who taught me loads of hands-on math activities. I wish I could remember the name of my Methods And Materials For Math professor, who was the first math teacher to help me *really* understand the Base 10 Number System, as well as a variety of totally awesome multiplication and division algorithms.

Table Of Contents

Note to teachers and parents..3
Developing the concept of multiplication..4
Multiplying Zero...10
Multiplying One...12
Multiplication Table...15
Multiplying Two (includes game)...17
Multiplying Three (includes game)...19
Multiplying Four (includes game)...27
Multiplying Five (includes game)..40
Multiplying Six (includes game)..44
Multiplying Seven (includes game)...49
Multiplying Eight (includes game)..53
Multiplying Nine (includes game)...59
Multiplying Ten (includes game)...70
Multiplying Eleven (includes game)..75
Multiplying Twelve (includes game)...80
What Is A MULTIPLE?...82
How To Figure Out If Any Number Is A Multiple Of Three.....................83
Multiples Of Six And Nine..85
Pattern/Logic Puzzles...86
Domino Multiplication (game)...96
Hands Up! (game)..97
The End Of The Grid (game)...98
Multiplication Table Bingo...101
Puzzle Pictures... 102
Multiplying Numbers Containing Zeros At The End............................109
What Is Division?...111
Multiplication And Division Fact Families.. ..113
The Answer Is Always 11 (calculation trick)..124
Odd Man Out (game)..125
Remainders..129
Dice Division (game)..132
A Superstar Multiplication Trick (mental multiplication)....................133
Multiplying 2-Digit By 1-Digit Numbers On Paper...............................136
Solve It! (game)..140
How To Multiply Two 2-Digit Numbers...142
The Standard Multiplication Algorithm...146
Multiplying Even LARGER Numbers..147
Dividing By 10, 100, And 1,000..149
Division And Decimals...151
Box Division...153
Division By Chunking...157
The Standard Division Algorithm..164
Congratulations Page..172

Note to teachers and parents (IMPORTANT – PLEASE READ!):

I wrote this book so that if a homeschooling parent wanted to, she could use it as a stand-alone curriculum for teaching multiplication and division. If a child is ready to, he can progress through the whole book without stopping.

If you begin with this book with a child under the age of nine or ten, however, chances are good that he may only be able to go so far (say, up to multiplying 12). At the point where the more advanced multiplication and division began to get confusing for him, feel free to take a break from this book and move on to another area of math for a while (such as geometry or measuring).

This book can also be used as part of a remedial math program for older children.

Whichever way you go, I suggest that you have the student(s) spend at least a week working to memorize each set of multiplication facts – 2X, 3X, 4X, etc. The games and puzzles at the beginning of the book will help them do just that in a fun manner. While they are working on memorizing the basic facts, give them word problems that require multiplication and division.

Classroom teachers – if you purchase this book with the intent on using it with the students in your classroom, you have permission to make as many copies of whatever sheets you would like for student use. HOWEVER, I ask that you respect all the hard work I put into producing this book by doing two things for me: first, by not making copies for other teachers. Any other teacher who wishes to use this book must purchase their own copy in order not to infringe on the copyright of this work. Second, please remember to respect the general copyright rules. If you end up teaching a workshop and decide to share any portion of this book with your colleagues, please tell your colleagues the title and author of this book, as well as the fact that they can purchase it from Amazon.

A bit of housekeeping: Whenever I have provided board games or game parts, I strongly suggest that either you or the student, after removing them from this book, glue them onto cardstock or poster board to increase the longevity of the games and pieces. You may also consider taking them to your local office supply store and laminating them.

ONE MORE THING: If you and your child(ren) find this book to be helpful and enjoyable, please take a moment to leave a positive review on the product page at Amazon.com. That way, more parents, teachers, and children who want to make multiplication fun will find this book more easily.

Thanks 1,000 X 1,000!

Emily Jacques
Former elementary school teacher, and homeschooling mom

P.S. – Throughout this book, I will be asking the student to discover various patterns. If they cannot figure it out, *please do not tell them!* The reason I do not provide the answers to my questions is that I want the students' brains to grow, for them to learn to think logically. If they can't see a pattern at this point in time, don't worry about it. Give them a few months, then revisit it and see how they do. Please do not take away the opportunity for brain growth in your child(ren)!

It's a fun word. Why? Well, for one thing, it contains *a lot* of other words. Check it out:

multiplication

Find at least 6 three-letter words in the word *multiplication*. Write them here:

_____   _____   _____

_____   _____   _____

_____   _____   _____

Find at least 4 four-letter words:

_____   _____   _____

_____   _____   _____

Sure, there are more! You may keep on going until your brain starts to leak, if you want. But when you're ready to get serious with math, turn the page, and get ready for some **FUN!**

There are three cats lying in your mother's flowerbed. They each have four legs. How many legs do all those cats have in total?

You can figure that answer out in several ways.

First, you can count the legs by ones: one, two, three, etc.

Second, you can count them by twos. That goes a little more quickly, but it's not the *quickest* way.

You could say to yourself, "Okay, I know 4+4=8, and 8+4=12." That's a good solution, too.

A fourth way would be to add the legs of each cat. There are three cats, so you would add four, three times. Like this: 4 + 4 + 4.

But the **EASIEST** way to solve this problem?

MULTIPLY!

4 x 3 = 12

See that X in between the 4 and the 3? That's called the "times", or "multiplication", sign. We read that number sentence like this: "Four times three equals twelve." That means, when you add 4 together three times (4 + 4 + 4) you get 12.

Four (legs) *three times* (because there are three cats) equals twelve. Get it? Four *times three* equals twelve.

So, what is multiplication?

If 5 X 2 (five times two) = 5 + 5,
and 7 X 4 = 7 + 7 + 7 + 7,
and 9 X 5 = 9 + 9 + 9 + 9 + 9…
        what is multiplication? (turn the page for the answer…)

Multiplication is *repeated addition.*

Read the above sentence five times. Now learn it backwards.

Just kidding!

Follow the patterns below to hammer the idea into your head that multiplication is repeated addition.

1 X 1 = 1
1 X 2 = 1 + 1
1 X 3 = 1 + 1 + 1
1 X 4 = 1 + 1 + 1 + 1
1 X 5 = __+__+__+__+__
1 X 6 =
2 X 2 = 2 + 2
2 X 3 = 2 + 2 + 2
2 X 4 =
2 X 5 =
2 X 6 =
2 X 7 =
3 X 2 = 3 + 3
3 X 3 = 3 + 3 + 3
3 X 4 =
3 X 5 =
3 X 6 =
3 X 7 =

And now, for two REALLY COOL ways to practice multiplication…

The first cool way to practice the concept of multiplication is by using number bars. You need graph paper to make the number bars. The last ten pages of this book each contain a blank grid. You will need six of those grids. The other four grids you may use however you like! You're welcome. ;)

Now, you're ready to make the number bars. Out of the first piece of graph paper, cut ten bars that each contain two squares, like this: ▢▢

Then, cut ten bars that each contain three squares, like this: ▢▢▢

Finally, cut ten bars that each contain four squares.

From the second sheet of graph paper, cut ten bars that each contain five squares, and five bars that each contain six squares.

From the third sheet, cut five bars that each contain six squares. Then cut ten bars that each contain seven squares.

From the other three sheets you will cut ten bars with eight squares each, ten with nine squares each, and ten with ten squares each.

That was a lot of work, right? But you're not quite done. You want all those number bars to last a long time, so what you need next is a big sheet of poster board (or you can use the sides of cereal/cracker/etc. boxes if you have enough saved up). Glue each number bar onto the poster board or cardstock, and cut them out again.

YOU'RE FINISHED! YIPPEE!

Go take a fruit snack break and come back when your hands are well-rested. ☺

Okay, are you ready to practice multiplication? Super! Here's how you do it.

3 X 3

See that? When you multiply using the number bars, you're going to make rectangles. The first number, or *factor*, in the number sentence (3 X 3) tells you **which** number bar to use – in this case, the 3 bar.

The second factor tells you **how many** 3 bars you need to put down.

To figure out the answer, simply count the little squares in the big rectangle. You should count 9.

Here's another example:

4 X 5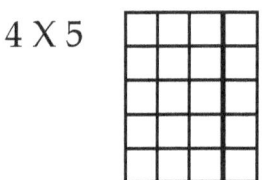

The first factor, 4, means you're going to use the 4 bar. The second factor, 5, means you're going to put 5 of the 4 bars down together. You will end up with a rectangle that is 4 squares wide and 5 squares long.

What is the *product* of 4 X 5? In other words, how many squares are there inside this rectangle?

The product of 4 X 5 is 20. We write a multiplication number sentence like this:
4 X 5 = 20.

Now I'm going to give you a bunch of other multiplication problems to solve. Use the number bars to figure out the product. Don't worry about writing down the answers right now; this activity is just to help you understand what multiplication is.

2 X 2   6 X 1   9 X 2   4 X 3   3 X 7   8 X 6   (Feel free to make up your own
                                                  multiplication problems using
5 X 3   2 X 4   6 X 4   3 X 6   8 X 5   7 X 6   the number bars.)

Okay, now you know the first REALLY COOL way to practice multiplication.

Here's the second way. You need:
- string, and
- a tiled floor (either real tile, or wood or linoleum tiles)

Let's call this activity, "Walk The Tiles."

To prepare for it, you need to cut 40 pieces of string that are the exact length of one floor tile.

The object is to mark out a rectangular **area** of the floor using the string, then count the tiles you mark off by walking on one tile at a time inside the rectangle.

Let's say you want to figure out what 7 X 2 is. Take seven pieces of string and lay them along the edge of seven floor tiles to form a straight line. Then take two pieces of string and lay them along the edge of two floor tiles, straight down from the end of the first of the seven strings you laid down. Here's what I mean:

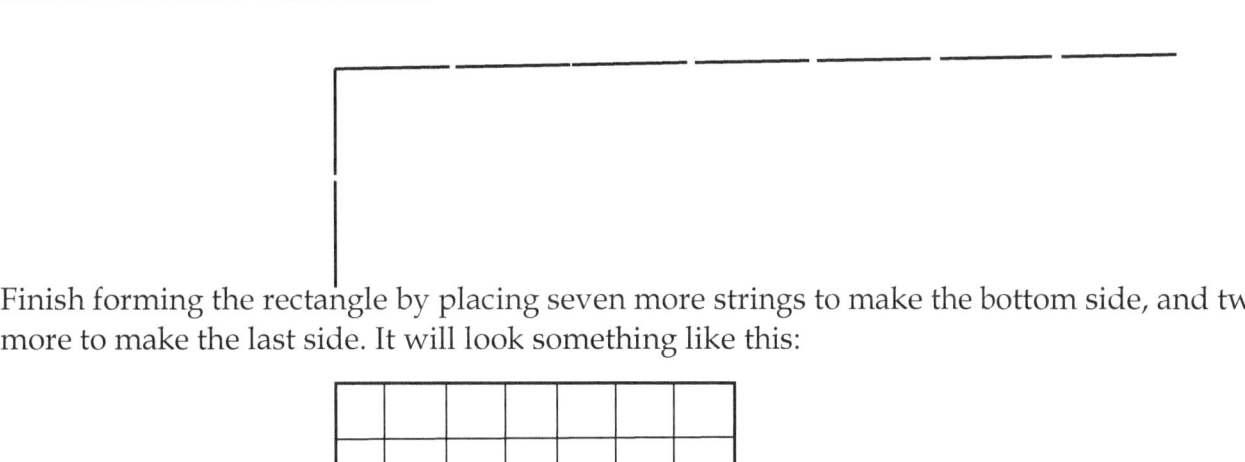

Finish forming the rectangle by placing seven more strings to make the bottom side, and two more to make the last side. It will look something like this:

Now you have a rectangle that is 7 floor tiles long by 2 floor tiles wide.

What is 7 X 2? Count the tiles inside the rectangle as you walk on them (careful not to knock the string askew!). You should count 14 floor tiles.

All right, time to get started. Grab a partner to work with you and work out some multiplication problems using the string and the floor.

Try starting easy with problems like 1 X 2 and 2 X 3. Then work your way up to larger rectangles like 5 X 9 and 7 X 7.   **HAVE FUN!** ☺

Multiplying ZERO

PSST! I have a question for you. Come a little closer…that's better.

Now, listen carefully. If I put two elephants in your bedroom, **ZERO TIMES**, how many elephants would there be?

I hope you answered **ZERO**. I did it zero times, so there can't be any elephants, right?

How about if I put 5 elephants in your bedroom zero times? How many elephants would there be then?

What about ten? I'm going to put TEN elephants in your bedroom ZERO times. How many elephants will be in your bedroom?

Watch out – I'm going to get **REALLY CRAZY**. What if I put 100 elephants in your bedroom zero times?

That's right – the answer would still be zero! If I'm putting any number of elephants ZERO times, there won't be any elephants, right?

And that's how easy it is to multiply zero.

Any number times zero equals zero.

Two elephants zero times = 2 X 0 = 0.

100 elephants zero times = 100 X 0 = 0.

To get used to that idea, follow the pattern below and fill in the missing numbers.

4 X 0 = 0

5 X 0 = 0

6 X 0 = 0

7 X 0 = ____

8 X 0 = ____

9 X 0 = ____

10 X 0 = ____

11 X ____ = 0

12 X 0 = ____

25 X 0 = ____

67 X ____ = 0

92 X ____ = 0

23,572 X 0 = ____

MOST EXCELLENT!

You've got the rule for multiplying zero, right? What is it?

Any number times zero equals zero.

You can't get much simpler than that…unless you're talking about multiplying one. Let's do that next!

Multiplying One

Remind me again what multiplication means?

Multiplication is repeated addition.

All right then, so tell me the answer to this multiplication problem: 1 X 8 =

What that means is that we are adding 8 just one time. That's 8 + ... uh, well, just 8, right? Because if we wrote down 8 + 8, that would be 8 *two* times. But we want to multiply it just *one* time.

Here's another way to look at it. Grab one of those eight bars you made a couple of pages ago. You know, a strip of paper that has eight little squares on it. Put the bar down one time. How many little squares are there?

You're right, eight! If we put eight squares down, one time, then we have eight squares.

1 X 8 = 8

Say you have five puppies in a kennel. You put that one kennel with five puppies on the kitchen floor. How many puppies are there?

Yep, five! 1 X 5 = 5

How about if I give you a dozen eggs, just one time? A dozen is twelve. So we're talking 12 X 1 = ??

Yes! 12 X 1 = 12.

Get the picture? What's the rule for multiplying one? (*Give me an answer before you read it on the next page, or I'll have to come to your house and tickle you!*)

Here's the rule for multiplying one:

any number times one equals that same number.

Follow the pattern below to drive that rule into your head.

1 X 1 = 1   1 X 4 = ___   1 X ___ = 15

1 X 2 = 2   1 X 5 = ___   1 X ___ = 31

1 X 3 = 3   1 X 6 = ___   1 X ___ = 76

1 X 5,678,214 = _____

Say the rule for multiplying 1 with me again:

any number times one equals that same number.

Just for the fun of it – what's the rule for multiplying zero?

Awesome! If you think you've got the rules for multiplying 0 and 1 down, let's move on to the part you've been waiting for:

THE GAMES!

Oh, wait, hold on one second. You don't know all the basic multiplication facts by heart yet, do you? Okay, no problem. I've got that covered.

On the next page is what people commonly call the "Times Table." A *real* mathematician calls it the "Multiplication Table", or "Multiplication Chart." You can use that table to figure out the answer to any of the basic multiplication facts through twelve.

Here's how it works. Say you want to know the product of 5 X 6. Put your index finger of your right hand on the 5 on the very top row of the table. Put your other index finger on the 6 that is down the very first *column*. Run your right finger straight down from the 5 as you run your left finger straight across from the 6.

They will meet at the number 30. And that's your answer. The product of the two factors 5 and 6 is 30. 5 X 6 = 30

So when you're playing the games and need to figure out an answer, just use the Multiplication Table and

ROCK ON, DUDE! ☺

Multiplication Table

×	1	2	3	4	5	6	7	8	9	10	11	12
1	1	2	3	4	5	6	7	8	9	10	11	12
2	2	4	6	8	10	12	14	16	18	20	22	24
3	3	6	9	12	15	18	21	24	27	30	33	36
4	4	8	12	16	20	24	28	32	36	40	44	48
5	5	10	15	20	25	30	35	40	45	50	55	60
6	6	12	18	24	30	36	42	48	54	60	66	72
7	7	14	21	28	35	42	49	56	63	70	77	84
8	8	16	24	32	40	48	56	64	72	80	88	96
9	9	18	27	36	45	54	63	72	81	90	99	108
10	10	20	30	40	50	60	70	80	90	100	110	120
11	11	22	33	44	55	66	77	88	99	110	121	132
12	12	24	36	48	60	72	84	96	108	120	132	144

Multiplying Two

Remember what multiplication is?

What? I can't hear you!

There you go!

Multiplication is repeated addition.

So then, what's another way to say 2 X 3?

2 X 3 = 3 + 3. Which equals…6! Right!

2 X 4 = 4 + 4 = …..what? Yes! Eight!

2 X 5 = 5 + 5 = _____

2 X 6 = 6 + 6 = _____

2 X 7 = 7 + 7 = _____

2 X 8 = ___ + ___ = _____

2 X 9 = ___ + ___ = _____

What do you notice about multiplying a number by two? (HINT: I'm looking for something about the addends in those addition problems…)

Did you say that the X2 problems are the same as addition doubles? Great job! Knowing that, you can probably play the very first multiplication game without even using the Multiplication Table.

NOTE: Starting with the X3 game and through the X9 game, after each game there will be a puzzle to help you memorize the facts for that number. Carefully cut out the puzzle page, then cut the pieces out on the dark line. Glue each piece separately onto poster board, cardstock, or chipboard (the cardboard that cereal boxes are made of), then cut out those pieces. As you re-assemble the puzzle pieces, say the multiplication fact you are putting together, along with the answer.

Times Two Slapjack

This game is not for the faint of heart. Or, shall I say, the slow-handed.

Players: 2

Materials needed:
- A deck of cards
- A very fast hand

Objective: To collect as many cards as you can

How to play

#1. Remove any jokers from the deck, or assign a number value to them. Queens and kings are worth 11.

#2. Shuffle the deck of cards and place it face down between the two players.

#3. First player draws a card and puts it face up *without* looking at it as he draws.

#4. Both players must multiply the number on the card by two in their heads.

#5. Whoever gets the product first slaps his hand over the card while saying the answer. If both players say the product simultaneously, whoever slaps the card first wins the round.

#6. The first player to say the product and slap the card gets to keep the card.

#7. The second player draws a card as the first player did. Repeat steps 4 – 5.

#8. Players take turns drawing each card.

#9. When a jack is drawn, no multiplication problem needs to be done. Whoever slaps the jack first gets to keep it.

#10. Play until all the cards in the stack are gone.

Multiplying Three – Three's Company

Players: 2

Materials needed:
- Board game on next page (color it however you want, and add some scenery such as a forest, town, park, etc.)
- Triangles, cut out and glued to construction paper or poster board. They are on the page after board game
- Two small objects to act as players' markers; for example, a dime and a penny or a black checker and a red checker
- A coin

Object of game: Accrue the most triangles by the time one of the players reaches the end.

How to play

#1. Place the triangle cut-outs in a stack between the two players.

#2. Player who is starting tosses coin and catches it. If it lands "heads" side up, she will move her marker two spaces. "Tails" side up, she will move one space.

#3. Whatever number she lands on, she must multiply it by three and give the correct product. For example, if she were to land on the number ten, she would say, "Thirty," because 3 X 10 = 30.

#4. If a player lands on a triangle on the board that has a multiple of three on it (3, 6, 9) **_and_** she correctly answers the multiplication problem, she collects a triangle from the stack.

#5. Player cannot reach "End" until and unless she flips the coin to give her the exact number to move to the "End" triangle.

#6. As soon as one player reaches "End," both players count their triangles. Whoever has the most triangles is the winner.

Three's Company

START

1 2 3 4 5 6 7 8 9

6 5 4 3 2 1 9 8 7

8 9 1 2 3 4 5 6 7 8

END

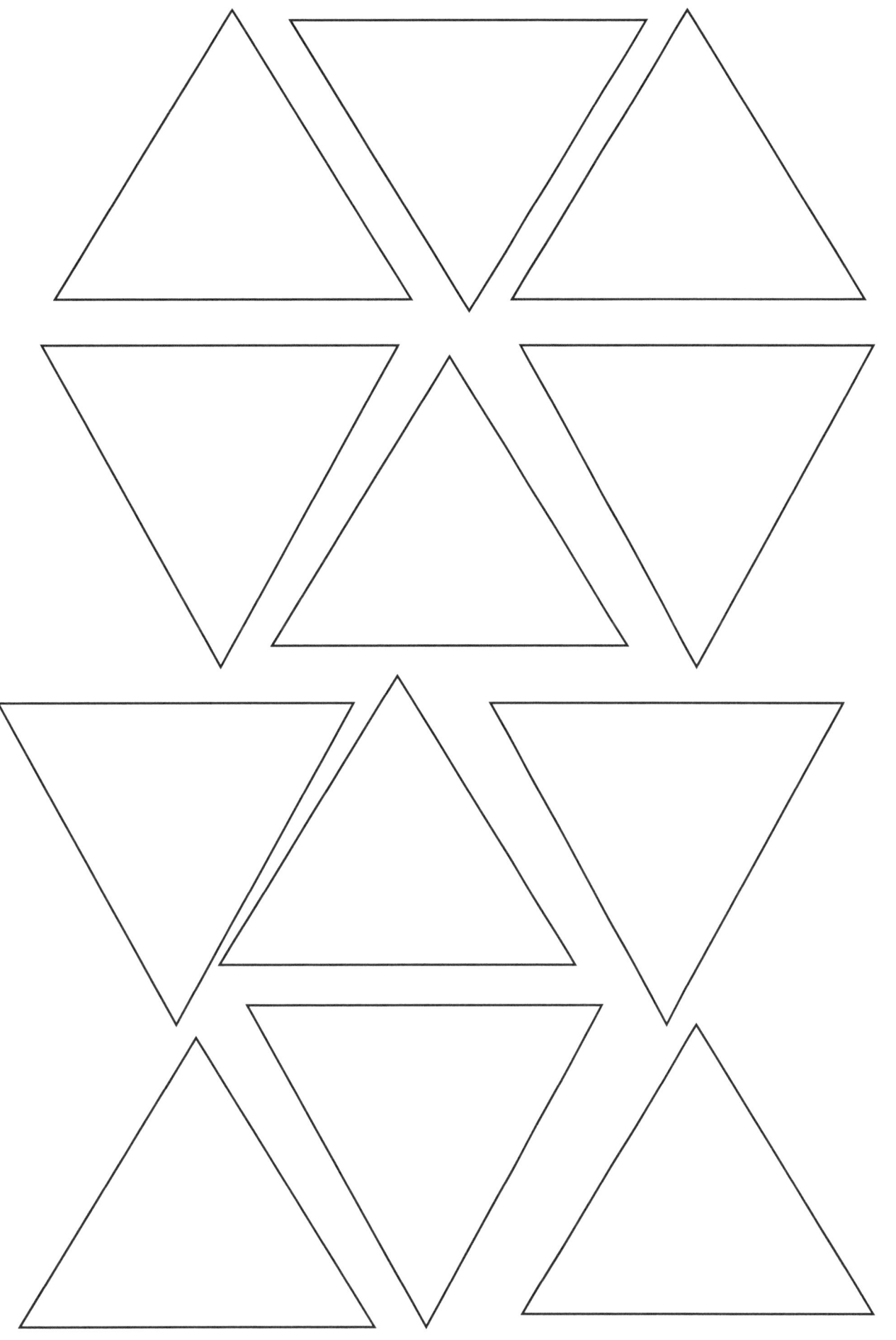

3 X 1 =3
3 X 2 =6
3 X 3 =9
3 X 4 =12
3 X 5 =15
3 X 6 =18
3 X 7 =21
3 X 8 =24
3 X 9 =27

Multiplying Four – Multiplication Bingo

Players: 2-4

Materials needed:
- Bingo boards (see next 4 pages)
- 16 markers (such as squares of paper) per player to cover numbers on board
- 9 strips of paper, each with one of the 4X facts written on it (4X1, 4X2, etc.)
- Multiplication table to find answers, if necessary
- Someone to be the "caller"

Object of game: Be the first player to cover either four numbers on his board in a row – across, diagonally, or down – or to cover all of his numbers, whichever way the players agree upon.

How to play

#1: Select a board and get 16 markers while the caller puts the multiplication facts in a box or small storage container.

#2: Caller pulls out one fact at a time and reads it aloud to the players.

#3: Players look on their boards to see if the answer is there. If so, they cover the number with a marker.

#4: Caller leaves fact out of container and pulls a new one out. When all the facts have been used up, caller puts them back in the container and starts over.

#5: As soon as a player has four numbers in a row covered (or the entire board, if this is what has been agreed upon), he calls, "Bingo!"

#6: Caller checks to make sure the covered numbers are the answers to the facts he has read. If so, the player who says, "Bingo!" wins.

x4 BINGO

12	36	8	28
8	48	20	24
24	4	48	32
40	32	16	44

x4 BINGO

16	44	4	36
24	4	40	24
20	12	48	28
32	36	28	8

x4 BINGO

44	12	20	40
8	32	4	24
40	20	48	16
28	44	12	36

x4 BINGO

4	20	44	12
8	16	32	4
12	8	24	40
28	36	48	16

4 X 1	=4	☆☆☆☆
4 X 2	=8	☆☆☆☆ ☆☆☆☆
4 X 3	=12	☆☆☆☆ ☆☆☆☆ ☆☆☆☆
4 X 4	=16	☆☆☆☆ ☆☆☆☆ ☆☆☆☆ ☆☆☆☆
4 X 5	=20	☆☆☆☆☆ ☆☆☆☆☆ ☆☆☆☆☆ ☆☆☆☆☆
4 X 6	=24	☆☆☆☆☆ ☆☆☆☆☆ ☆☆☆☆☆ ☆☆☆☆☆
4 X 7	=28	☆☆☆☆☆ ☆☆☆☆☆ ☆☆☆☆☆ ☆☆☆☆☆
4 X 8	=32	☆☆☆☆ ☆☆☆☆ ☆☆☆☆ ☆☆☆☆
4 X 9	=36	☆☆☆☆ ☆☆☆☆ ☆☆☆☆ ☆☆☆☆

A REALLY FUN Pattern!

Hold on. I know you're eager to move on to multiplying by five. But what if I told you that there is a super-fun pattern when you multiply numbers that only have the digit four?

To see what I mean, go get a calculator . Go on. I'll wait.

Got it? Okay. Use the calculator to find the answers to the following multiplication problems. Write down the answers in the blanks below.

4 X 4 = _____

44 X 4 = _____

444 X 4 = _____

4,444 X 4 = _____

You see the pattern? Neat, huh? And you're so smart, I'll bet you can write down the answer to the next problem without even using your calculator!

44,444 X 4 = _____

And now, on to multiplying by five!

Multiplying Five – No More Nickels

Usually, people want to earn *more* money. But the object of this game is to get rid of it all!

Players: 2

Materials needed:
- 40 nickels, 20 per player
- Two sets of number cards 1-9 (can use playing cards, or write them on index cards). In other words, you need two cards with the number 1, two with the number 2, and so on.

Object of the game: Be the first to get rid of your entire stack of nickels.

How to play

#1: Each player stacks his nickels, such as two stacks of ten or four stacks of five.

#2: Shuffle the cards and put them face-down between the two players.

#3: The first player draws the top card from the stack. He must multiply that number times 5 and say the product. Then he removes whatever number of nickels from his stack that the card says. For example, if he draws the number 2, he says, "Five times two equals ten." Then he takes two nickels from the stack and sets them aside.

#4: Players take turns drawing cards, saying the corresponding multiplication fact ("Five times…"), and discarding nickels.

#5: Game ends when one player has no more nickels left. She must draw the exact right card to be able to discard the last of her nickels.

5 X 1	=5	⬠
5 X 2	=10	⬠ ⬠
5 X 3	=15	⬠ ⬠ ⬠
5 X 4	=20	⬠ ⬠ ⬠ ⬠
5 X 5	=25	⬠ ⬠ ⬠ ⬠ ⬠
5 X 6	=30	⬠ ⬠ ⬠ ⬠ ⬠ ⬠
5 X 7	=35	⬠ ⬠ ⬠ ⬠ ⬠ ⬠ ⬠
5 X 8	=40	⬠ ⬠ ⬠ ⬠ ⬠ ⬠ ⬠ ⬠
5 x 9	=45	⬠ ⬠ ⬠ ⬠ ⬠ ⬠ ⬠ ⬠ ⬠

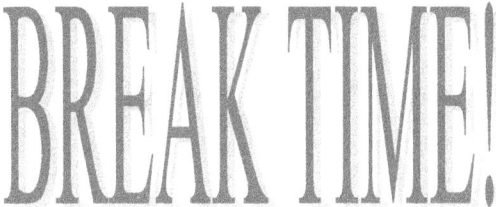

Time to talk about a **REALLY BIG WORD!**

Okay, so maybe it's not *that* big of a word. And it's really two words. Here they are:

commutative property

Heard that before? You may have, back when you were learning how to add and subtract. The word "commutative" is related to the word "community," and it simply means that when you add numbers together, you will get the same answer no matter what order you put them in.

For example, 1 + 3 = 4, and 3 + 1 = 4

    5 + 3 + 6 = 14, and 6 + 5 + 3 = 14.

Mathematicians call this the "commutative property of addition."

Well, guess what? There is also a commutative property of multiplication! You're so smart, I'll bet you can tell me what that means before I even tell you!

Just like with addition, the order in which you put the numbers when you are multiplying them together doesn't matter.

So if 2 X 3 = 6, then 3 X 2 = 6.
   If 4 X 5 = 20, then 5 X 4 = _____.
   If 6 X 4 = 24, then 4 X 6 = _____.

Do you know what that means? That means that you don't have to memorize nearly as many multiplication facts as you thought! If you know the product of 3 X 6, then you know the product of 6 X 3.

Isn't that fun?

Okay, back to our regularly scheduled multiplication page...

Multiplying Six – Silly Sixes

Be warned: you need to "get your silly on" before you play this game!

Players: 2-3

Materials needed:

- 9 sheets of paper (8.5 by 11 inches), cut in half so that you have 18 total sheets
- Slips of paper with silly instructions on them; see next page

Object of game: This game is a fun way to practice the 6X facts, rather than a competition. Aim to have each player have five turns before calling it quits.

How to play

#1: On 9 of the sheets of paper, write the **ANSWER** to one of the 6X facts. So one paper will have "6", the next "12" (for 6X2), the next "18" (for 6X3), and so on. Repeat for the other 9 sheets of paper.

#2: Arrange the papers on the floor in two equal rows, about one foot apart.

#3: Place the slips of paper with silly instructions inside a small box or plastic storage container (things like clean yogurt cups work well).

#4: Player one draws a slip of paper from the container. He must accomplish two things at the same time: first, make the silly action, and second, walk on one numbered paper at a time. So if the card says "Pat your belly", he must pat his belly at the same time he steps onto the different sheets of paper. **Go slowly** so that you don't slip and fall!

#5: While the first player is doing this, the other player counts to five as slowly or as quickly as she wants (within reason).

#6: When the second player says, "Five," the first player stops. Then he looks down to see where his right foot has landed (it should be on a numbered piece of paper). He has to say the 6X fact that goes with the number. For example, if the number on the paper is 12, he has to say, "Six times two."

#7: Player 2 decides if player 1 is correct. If player 1 gets the fact right, he can walk back to the starting place normally. If he gets it wrong, he has to do the silly action on his way back to the starting place.

#8: Repeat steps 4-7 with the second player. Continue game until each player has had at least five turns.

Silly Actions

Quack like a duck   Flap your elbows like bird wings

Walk while twisting your hips   Make a funny face

Squat down really low every time you take a step

Pat your belly with both hands while you walk

Scratch your head with one hand and pat your thigh with the other

Keep opening and closing your mouth. When you open your mouth, open it as wide as you can.

Jump on both feet while clapping

Wiggle your hands behind your bottom like you have tail feathers

6 X 1	=6
6 X 2	=12
6 X 3	=18
6 X 4	=24
6 X 5	=30
6 X 6	=36
6 X 7	=42
6 X 8	=48
6 X 9	=54

Multiplying Seven – Name That Factor!

Just like Silly Sixes, with this game you have to give the multiplication factors for the product that you get.

Players: 2

Materials needed:
Cards (index is fine) with the answers to the 7X (through 7X9) written on them: 7, 14, 21, 28, 35, 42, 49, 56, 63, 70

Object of game: Have the most cards by the time the stack is depleted.

How to play

#1: Shuffle the cards well. Place them face-down between the two players.

#2: Player 1 draws a card.

#3: As soon as Player 1 looks at number on card, Player 2 quietly counts, "One one-thousand, two one-thousand…" to "five one-thousand". Player 1 must give the correct 7X multiplication sentence before Player 2 finishes counting. For example, if the card has the number 77 on it, Player 1 has five seconds to say, "Seven times eleven!"

#4: If Player 1 gives correct answer in the allotted time, she gets to keep the card.

#5: Player 2 then takes his turn while Player 1 counts.

#6: Game ends when all ten cards are gone from the stack. Whoever has the most cards, wins.

LISTEN UP:
As the players get more proficient with giving the correct 7X multiplication facts, shorten the time frame in which they can give an answer.

7 X 1	=7	✿✿✿✿✿✿✿
7 X 2	=14	✿✿✿✿✿✿✿✿✿✿✿✿✿✿
7 X 3	=21	✿✿✿✿✿✿✿✿✿✿✿✿✿✿✿✿✿✿✿✿✿
7 X 4	= 28	✿✿✿✿✿✿✿✿✿✿✿✿✿✿✿✿✿✿✿✿✿✿✿✿✿✿✿✿
7 X 5	=35	✿✿✿✿✿✿✿✿✿✿✿✿✿✿✿✿✿✿✿✿✿✿✿✿✿✿✿✿✿✿✿✿✿✿✿
7 X 6	=42	✿✿
7 X 7	=49	✿✿✿
7 X 8	=56	✿✿
7 X 9	=63	✿✿✿

Multiplying Eight – Crazy Eight

No, this isn't the card game. But, who knows? It might just be more fun!

Players: 2-3
Materials needed:
- Crazy Eight board (see next page)
- Markers (such as buttons or coins), one per player
- Die
- Deck of playing cards with all the jokers and face cards removed (or make your own set of cards with the numbers 1-9 written on them, each number written on three to four cards)

Object of game: Be the first player to reach the innermost circle of the board.

How to play
#1: Shuffle cards and set them face-down between the players.

#2: Each player puts his marker on "Start."

#3: Player 1 draws a card from the deck, then multiplies that number by 8 and gives the product.

#4: The other players check the answer against the multiplication table to make sure the answer is correct.

#5: If Player 1 has given the correct answer, he throws the die and moves his marker in any direction the number of spaces that the die shows. The first move out of "Start" is counted as space number one.

#6: If he lands on a space with an opening in the circle, he may move through the opening to get closer to "End." The next player will take a turn.

#7: If Player 1 gives an incorrect answer, his turn is over.

#8: The next player takes a turn. Repeat steps #3-#7.

#9: Game ends when a player reaches the center of the circle labeled "END".

8 X 1	=8	
8 X 2	=16	
8 X 3	=24	
8 X 4	=32	
8 X 5	=40	
8 X 6	=48	
8 X 7	=56	
8 X 8	=64	
8 X 9	=72	

Multiplying Nine

I know you **REALLY WANT** to get to the game to help you learn the X9 facts. But what if I told you that I have a secret that might help you learn the facts even *faster*?

The secret is the **REALLY FUN** pattern of the products of multiplying nine. Grab your multiplication table and write in the answers to the facts listed below. I'll talk to you when you get to the end!

P.S. – **Make sure you line up the units place and tens place** of all the answers. This is important.

9 X 1 = _____

9 X 2 = _____

9 X 3 = _____

9 X 4 = _____

9 X 5 = _____

9 X 6 = _____

9 X 7 = _____

9 X 8 = _____

9 X 9 = _____

9 X 10 = _____

All done? **Great!** ☺

Now, go back to the previous page and put your finger on the first answer at the very top, for 9 X 1. Say that product. Go to the answer for 9 X 2. Say the digit that is in the *units* (or ones) place. You should have said, "Eight." Keep going down. Say the digit that is in the units place for each product, in order, all the way down. Come back here when you're done.

Got it? Great! What can you tell me about the digits in the units place for the X9 facts?

Now do the exact same thing for the tens place. The digit in the *tens* place for 9 X 1 is 0, so go ahead and say zero. Which digit is in the *tens* place for 9 X 2? Yes, the digit one. Keep going down and saying the digit that is in the tens place for each product. Come back here when you've finished.

What did you notice about the digits in the tens place for the X9 facts?

So if you remember the pattern for both the units and the tens place, it can help you figure out a 9X fact you don't yet have memorized.

Ready for the X9 game now? Then, let's go!

Nix Nine Animals

Players: 2-4

Materials needed:
- Deck of playing cards with all the face cards removed, OR set of cards with 1-9 written them, so that each number is written on three cards.
- One Nix Nine Animal card per player (see next page)
- Nine markers (buttons, coins, squares of paper, etc.) per player (what you used for the X4 Bingo game is fine)

Object of game: Be the first player to cover all animals on your board with a marker.

How to play

#1: Shuffle the cards and place them face down between the players.

#2: Each player selects a Nix Nine Animal card and collects nine markers.

#3: Player 1 draws the top card from the stack. Whatever number she draws, she multiplies it by 9. If the answer is an even number, her turn ends. If the answer is an odd number, she may cover one of the animal pictures on the card with a marker.

#4: Player 2 takes a turn just as Player 1 did.

#5: Players continue to take turns drawing cards and solving the X9 problems.

#6: Game ends as soon as one of the players has covered all of the animal pictures on her card.

Nix Nine Animal Cards

9 X 1	=9	♡♡♡♡♡♡♡♡♡
9 X 2	=18	♡♡♡♡♡♡♡♡♡ ♡♡♡♡♡♡♡♡♡
9 X 3	=27	♡♡♡♡♡♡♡♡♡ ♡♡♡♡♡♡♡♡♡ ♡♡♡♡♡♡♡♡♡
9 X 4	=36	♡♡♡♡♡♡♡♡♡ × 4
9 X 5	=45	♡♡♡♡♡♡♡♡♡ × 5
9 X 6	=54	♡♡♡♡♡♡♡♡♡ × 6
9 X 7	=63	♡♡♡♡♡♡♡♡♡ × 7
9 X 8	=72	♡♡♡♡♡♡♡♡♡ × 8
9 X 9	=81	♡♡♡♡♡♡♡♡♡ × 9

A "Handy" Trick For Practicing The Nine Facts

Got two hands? Ten fingers?* If so, then here's a fun and "handy" way to practice the nine multiplication facts.

First, open both hands and hold them in front of you so that the palms face your face.

Next, starting with the thumb of the left hand, imagine that all your fingers are numbered from one to ten. The thumb on your left hand will be number one, and the thumb on your right hand will be number ten.

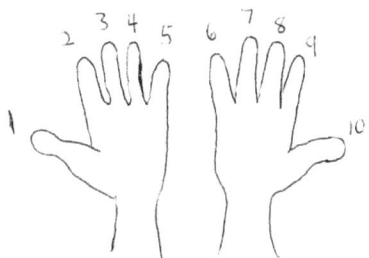

After that, choose any of the 9X facts, from 9x1 to 9x9. What is the number you are multiplying by 9? Count your fingers, and put down the corresponding finger, bending it toward your palm.

For example, say you want to solve 9 X 6. Which finger is number 6? The pinky of your right hand. The problem is 9 X 6, so you bend down that finger, finger number 6.

Now, how many fingers are up on the left side of the bent finger? Five, right? So 5 will be the digit in the tens place of the answer.

How many fingers are up *after* the bent finger, to the right of it? You should count four. So 4 will be the digit in the units place of the answer.

5 and 4. 9 X 6 = 54.

Is that an awesome way to practice your 9X facts, or what?

9 × 6 = 54.

Put down finger #6.
Then you have 5 to the left of it
and 4 to the right of it. 54. See?

*I'm not being silly here – some kids don't have two hands or all their fingers.

Multiplying Ten

Ten? **TEN?** Do I really expect you to multiply by ten? But that's a big number! Won't it be hard?

You tell me. I am going to start two patterns of multiplying numbers by ten. Continue the patterns, starting with the one on the left that begins with 10 X 1 = 10. After you have finished 10 X 10 at the bottom of the first column, go up to the second column and figure out that pattern. Then see if you can figure out the rule for multiplying numbers by ten.

10 X 1 = 10   10 X 17 = 170

10 X 2 = 20   10 X 23 = 230

10 X 3 = 30   10 X 45 = 450

10 X 4 = _____   10 X 51 = _____

10 X 5 = _____   10 X 64 = _____

10 X 6 = _____   10 X 87 = _____

10 X 7 = _____   10 X 93 = _____

10 X 8 = _____   10 X 453 = _____

10 X 9 = _____   10 X 872 = _____

10 X 10 = _____   10 X 1,352 = _____

Did you figure out the rule? Yes?

YOU ROCK, DUDE!

Fill in the missing words to explain the rule for multiplying ten: "When you multiply any number by ten, all you have to do is _____.

Fantastic! Let's move on to the next game, shall we?

Race To 100

Now that you know the rule for multiplying by ten, you shouldn't need your multiplication table to play this game, right?

RIGHT??

Players: 2

Materials needed:
- Race To 100 game board (see two pages from here)
- One game piece per player, such as coins, buttons, etc.
- 30 cards with the numbers 1-10 written on them, 3 cards for each number
- Scratch paper

Object of game: Get as close to 100 as you can after three turns, and have the highest number after three rounds.

How to play

#1: Shuffle cards and place them face down, place game board between both players.

#2: Players both set their markers on 0 on the game board.

#3: Player 1 draws the card on top and multiplies that number by 10.

#4: He moves his marker to the corresponding number on the board.

#5: Player 2 takes a turn, following steps #3 and #4.

#6: On his next turn, Player 1 must add the product to the number his marker is already on, and move his marker – if he can – to the new number.
For example, if Player 1's marker is on 30, and he draws a 4 on his second turn, he multiplies 4 by 10, which is 40. He then adds 30 and 40, which equals 70. So he moves his marker to 70.

#7: If, when a player adds the new product to the number where his marker is sitting and the answer is greater than 100, he cannot move. For example, if his marker is on 50, and he draws an 8, 8X10=80 and 80+50=130, which is greater than 100. So the player would keep his marker on the 50.

#8: Continue playing like this until each player has taken three turns (you might want to use tally marks to keep track of this). After each has had three turns, write down the final number where each player ended up on the board. This is the end of Round 1.

#9: Follow steps #3 - #8 for two more rounds.

#10: When all three rounds have been completed, add up the three numbers for each player. For example, if Player 1 had a 50 for Round 1, 70 for Round 2, and 60 for Round 3, his total would be 50 + 70 + 60 = 180.

#11: The player with the largest sum is the winner.

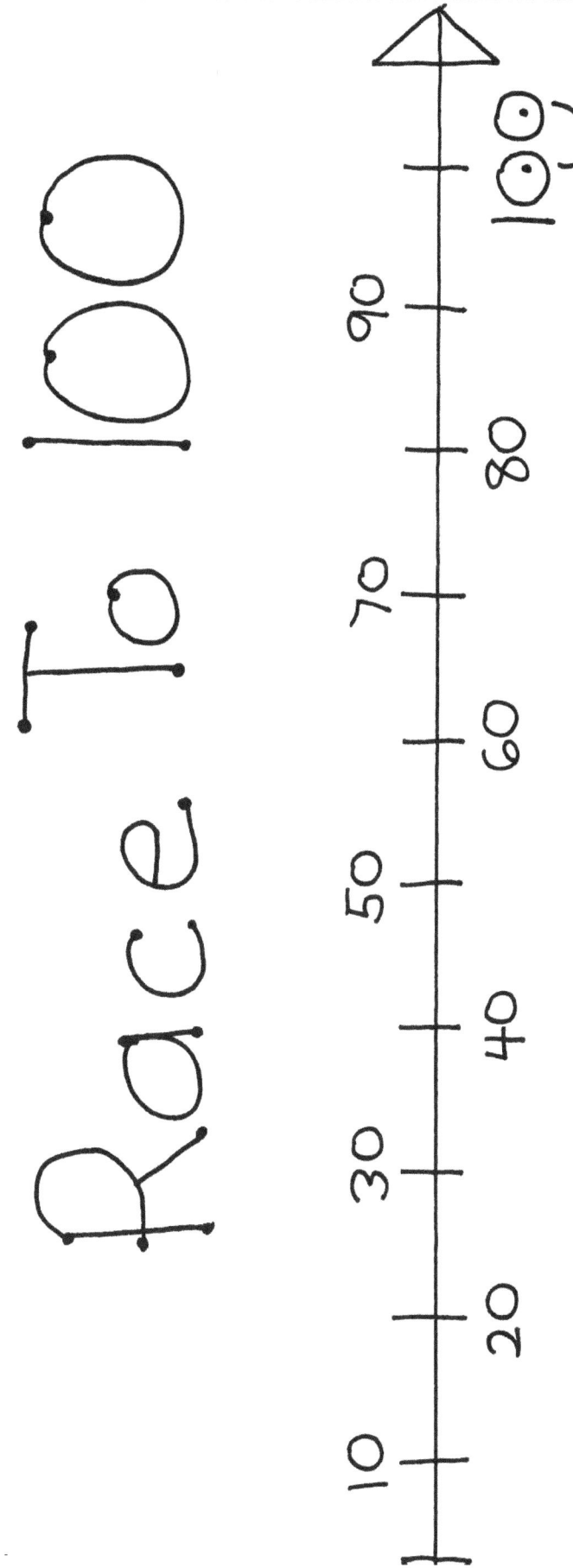

Multiplying Eleven

How about we take a break from games for a minute or two? I want to show you the *TOTALLY FUN* pattern for multiplying one-digit numbers by 11. I'll start it, and you finish it. Then, tell the rule for multiplying one-digit numbers by 11.

11 X 1 = 11   11 X 6 = _____

11 X 2 = 22   11 X 7 = _____

11 X 3 = 33   11 X 8 = _____

11 X 4 = _____   11 X 9 = _____

11 X 5 = _____

Got it? So, what's the rule for multiplying one-digit numbers by 11?

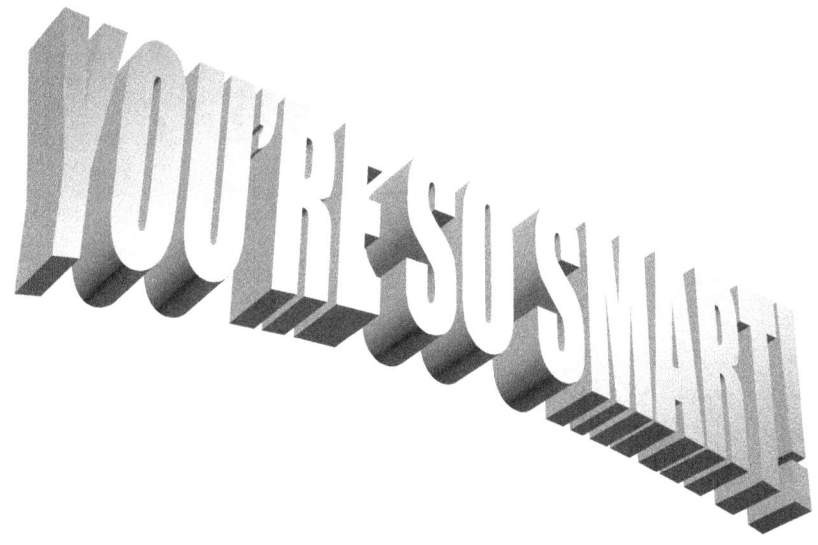

Later on, I'll tell you the trick for multiplying numbers with more than one digit by 11. For now, let's get to the game!

Remembering 11

Players: 2

Materials needed:
- 2 sets of cards with X11 facts and answers (see next page; if the cards are too small, copy them onto index cards). You MUST glue the cards onto poster board or cardboard so that players will not be able to read the figures on them when they are upside down.
- Blue and red crayons

Object of game: To collect the most cards

How to play

#1: With the blue crayon, color the corners of all the cards in one of the sets of cards blue.

#2: Color the other set the same way with the red crayon.

#3: Mix both sets of cards together well. Arrange them *face down* in three rows with three or four cards each on the floor or table.

#4: Player 1 turns over two cards of her choice. If they match, she gets to keep the cards. A match is the same color X11 fact and its corresponding answer. For example, a blue 11 X 2 and a blue 22 would be a match, but a blue 11 X 2 and a red 22 would *not* be a match. Neither would a blue 11 X 2 and a blue 33, because 11 X 2 does not equal 33.

#5: If the cards are not a match, she turns them back over in the same place and Player 2 takes a turn.

#6: Game ends when someone has picked up the last two cards.

11 X 1	11	11 X 1	11
11 X 2	22	11 X 2	22
11 X 3	33	11 X 3	33
11 X 4	44	11 X 4	44
11 X 5	55	11 X 5	55
11 X 6	66	11 X 6	66
11 X 7	77	11 X 7	77
11 X 8	88	11 X 8	88
11 X 9	99	11 X 9	99

Another SUPER FUN Pattern!

Remember back when you were multiplying by four, you learned a fun pattern that happens when you multiply numbers with only the digit four?

Well, a similar thing happens when you multiply numbers with only the digit 1. You end up with a totally fun pattern! To find the pattern, grab a calculator and solve the problems below. Write in the answers to see the pattern better.

    11 X 11 =_____

    111 X 11 =_____

    1,111 X 11 =_____

See it? Doesn't that totally rock? Now, show me how smart you are and write down the answer to the following problem *without* using a calculator:

    11,111 X 11 =_____

Feeling smart? Why not show an unsuspecting grown-up how smart you are, and tell them the answer to 111,111 X 11 without using your calculator? (Actually, if you tell them the answers to any of the other 11X problems above, they will probably be impressed.)

All right, time to move on to multiplying 12!

Multiplying Twelve

PATTERN TIME!

Yep, when you multiply consecutive numbers ("consecutive" means one after the other in order) by twelve, you get an interesting pattern.

Fill in the products below, then look at the units place, starting at the top answer and working your way down to the last one at the bottom.

12 X 1 =_____

12 X 2 =_____

12 X 3 =_____

12 X 4 =_____

12 X 5 =_____

12 X 6 =_____

12 X 7 =_____

12 X 8 =_____

12 X 9 =_____

12 X 10 =_____

What's the pattern you see in the units place? Why?

Well, what digit is in the units place in the number 12?

That's right, a 2! **Get it?**

All right! Now, we're ready for the X12 game!

The Dirty Dozen

Players: 2

Materials needed:

- 1 egg carton (with 12 sections)
- Sharpie or other permanent marker
- Coin
- Two buttons (or other items to use as game pieces that will fit inside an egg carton)

Object of the game: Be the first to reach the number 12 in the egg carton.

How to play

#1: Using the marker, number the bottom of each of the egg carton sections 1 – 12, in order.

#2: Use the edge of the egg carton nearest the section that is numbered 1 as "Start." Both players place their game pieces there. If they tend to fall off, just put them on the surface right below the egg carton.

#3: Decide on a time limit for players to solve multiplication problems, up to fifteen seconds.

#4: Player 1 flips a coin. Heads, he moves 1 space. Tails, he moves 2 spaces. He must follow the numbers in order.

#5: Player 1 must multiply by 12 whatever number he lands on. For example, if he lands on 2, he will state, "Twelve times two equals…" and give the answer *without using the multiplication table.*

#6: Player 2 checks the answer against the multiplication table. If Player 1 gives an incorrect answer, Player 2 will get two turns in a row.

#7: Repeat steps #3 - #5 with players taking turns.

#8: Game ends when first player lands on the 12 in the egg carton.

What Is A MULTIPLE?

Wow, you're really getting close to becoming a **MATH SUPERSTAR**, learning all those multiplication facts!

Want to get one step closer? Okay! Let's talk about *multiples*.

A **multiple** is the answer you get when you multiply a certain number. To begin, let's look at the first nine multiples of 2.

2 X 1 = 2   2 X 4 = 4   2 X 7 = 14
2 X 2 = 4   2 X 5 = 10   2 X 8 = 16
2 X 3 = 6   2 X 6 = 12   2 X 9 = 18

In that list, you see that the first nine multiples of 2 are 2, 4, 6, 8, 10, 12, 14, 16, and 18.

The first three multiples of 3 are 3, 6, 9, because 3 X 1 = 3, 3 X 2 = 6, and 3 X 3 = 9. See? Easy peasy!

Now, tell me in your own words…what is a **multiple**?

Great! Since you're so smart, now write down the first six multiples of 4:

_____ _____ _____ _____ _____ _____

This one will be easy. What are the first ten multiples of 5?

_____ _____ _____ _____ _____ _____

_____ _____ _____

FANTASTIC!   Now that you understand what a multiple is, let's talk about a tremendously super trick about the multiples of 3!

How To Figure Out If Any Number Is A Multiple Of Three

True or false: 52,625 is a multiple of 3. (**NO FAIR USING A CALCULATOR TO DIVIDE IT BY 3!!!**)

Not sure? At least without your calculator? Then, allow me to teach you a super-fun way to figure out if any given number – even HUMUNGOUS ones – is a multiple of 3.

The first step is to fill in the answers to all the X3 number sentences below.

The second step is to add together the digits in the answers that have more than one digit. For example, 3 X 4 = 12. 12 has two digits, so you add them together: 1 + 2 = 3.

Finally, look down the last column of numbers you create and figure out the pattern. Ready? Here we go!

3 X 1 = _____

3 X 2 = _____

3 X 3 = _____

3 X 4 = _____ (add those digits together) ___+___=_____

3 X 5 = _____ ___+___=_____

3 X 6 = _____ ___+___=_____

3 X 7 = _____ ___+___=_____

3 X 8 = _____ ___+___=_____

3 X 9 = _____ ___+___=_____

If you did it correctly, you should see the same 3 one-digit numbers repeated in the same order all the way down. You know that a number is a multiple of 3 if, when you add the digits in the number, the sum turns out to be one of those three numbers.

"But, hold on!" you protest. "When I add together the digits of 52,625, I get another multi-digit number!"

You're right! So, add *those* digits, then add *those* digits, and so on, until you come down to one digit. Let's do it together.

$$5 + 2 + 6 + 2 + 5 = 20, \quad 2 + 0 = 2$$

52,625 breaks down to 2. So, is it a multiple of three?

How about the number just after it, 52,626?

Is that a neat trick, or what?

Want to learn a couple of other multiple tricks? Okay, here we go!

Multiples Of Six And Nine

On the left side of this page are the X6 multiplication facts through 12. On the right are the X9 multiplication facts through 12.

Just like you did for the X3 facts, you are going to look for a pattern in the answers of the multiples. When you have a multiple that is two digits, you are going to add those digits together and write the number down. For example, 6 X 2 = 12. Twelve is a two-digit number, so you will add the digits together: 1 + 2 = 3.

Got it? Then, go ahead!

6 X 1 = _____   9 X 1 = _____

6 X 2 = _____   9 X 2 = _____

6 X 3 = _____   9 X 3 = _____

6 X 4 = _____   9 X 4 = _____

6 X 5 = _____   9 X 5 = _____

6 X 6 = _____   9 X 6 = _____

6 X 7 = _____   9 X 7 = _____

6 X 8 = _____   9 X 8 = _____

6 X 9 = _____   9 X 9 = _____

6 X 10 = _____   9 X 10 = _____

6 X 11 = _____   9 X 11 = _____

6 X 12 = _____   9 X 12 = _____

What pattern do you notice for the multiples of six?

What do you notice about the multiples of nine?

Doesn't that totally ROCK? ☺

Yes, you can find patterns for multiples of other numbers, too. I'll let you do that on your own. For now, let's move onto some really fun puzzles! HINT: there are missing numbers AND picture elements that you'll need to fill in.

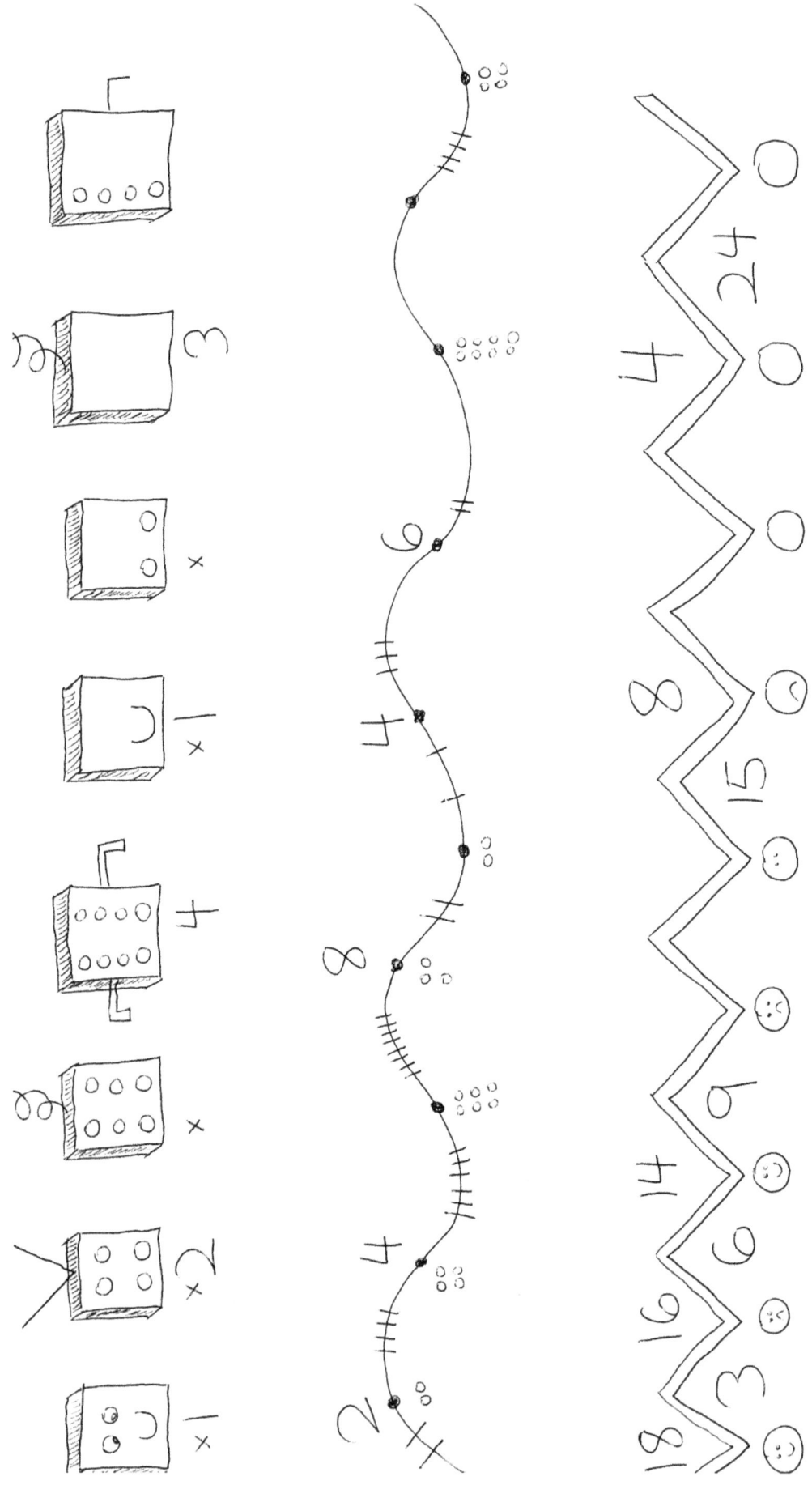

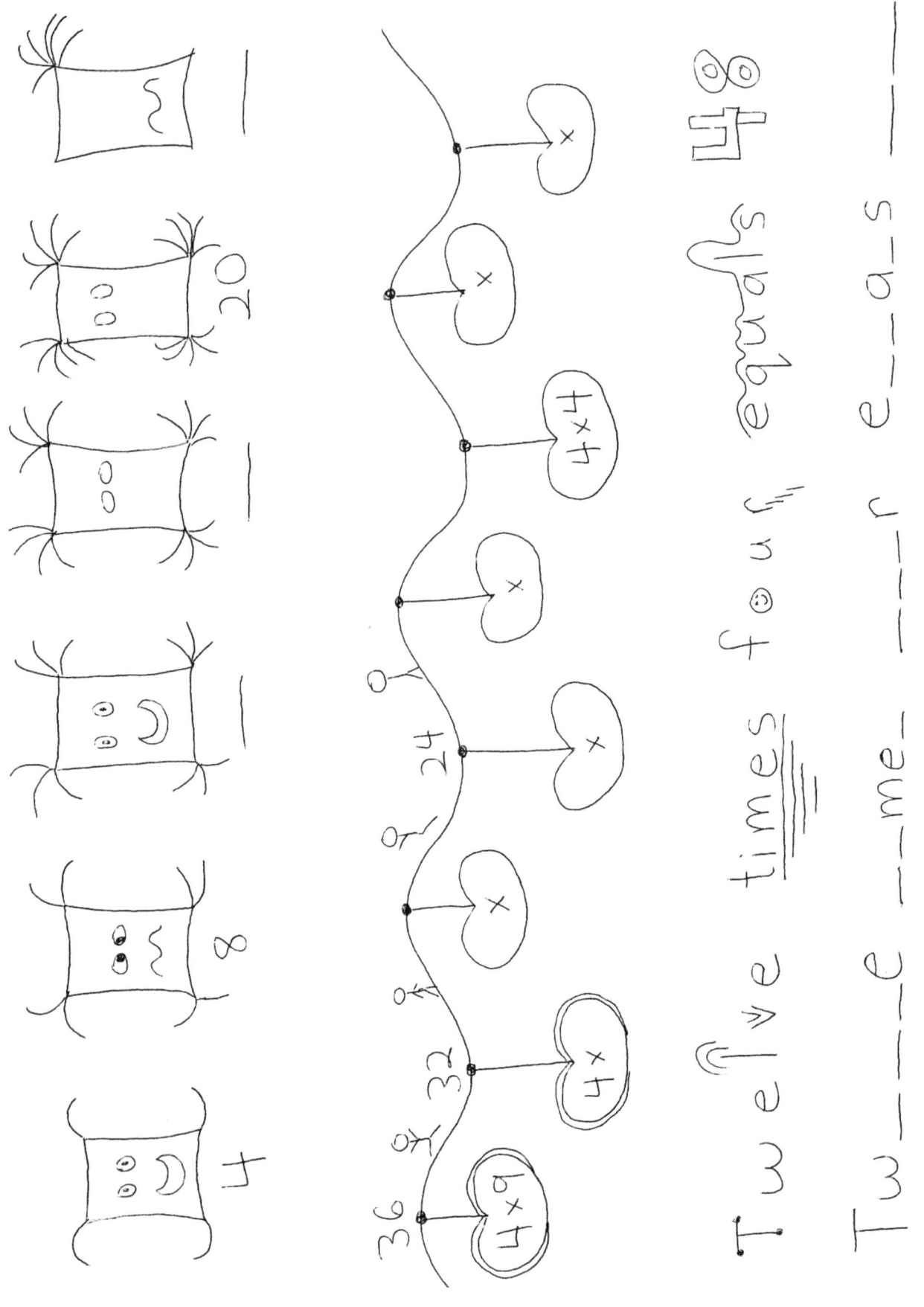

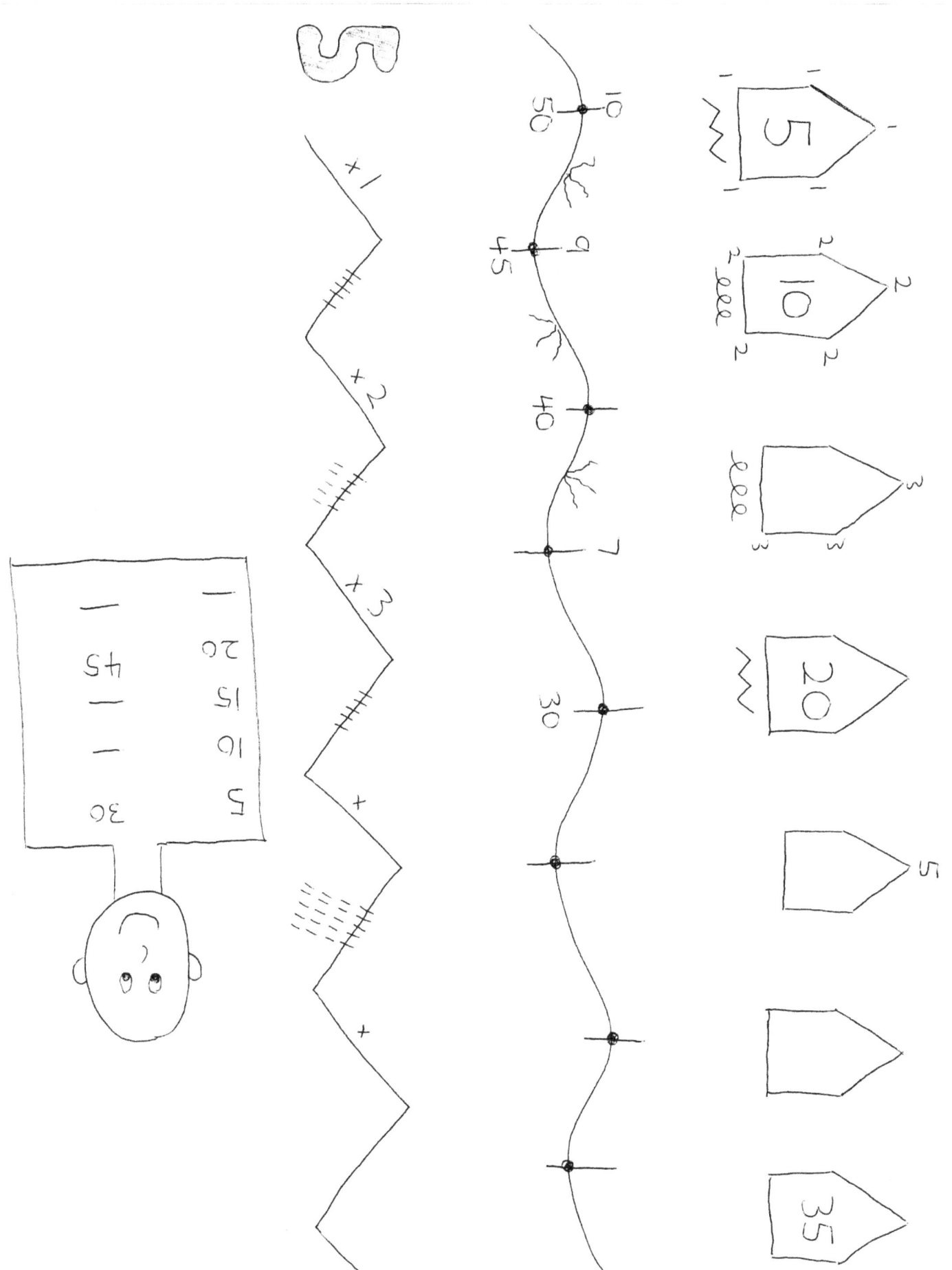

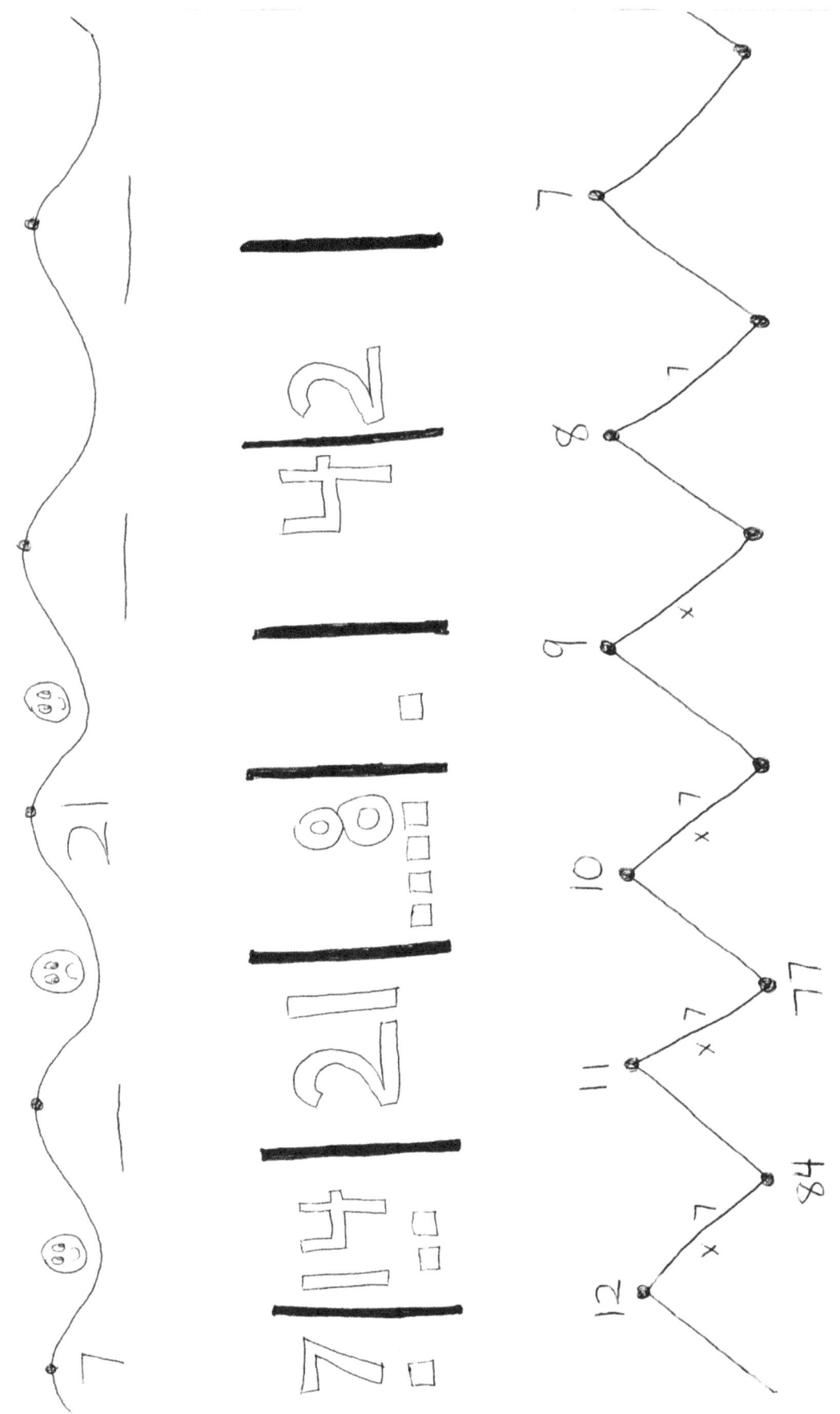

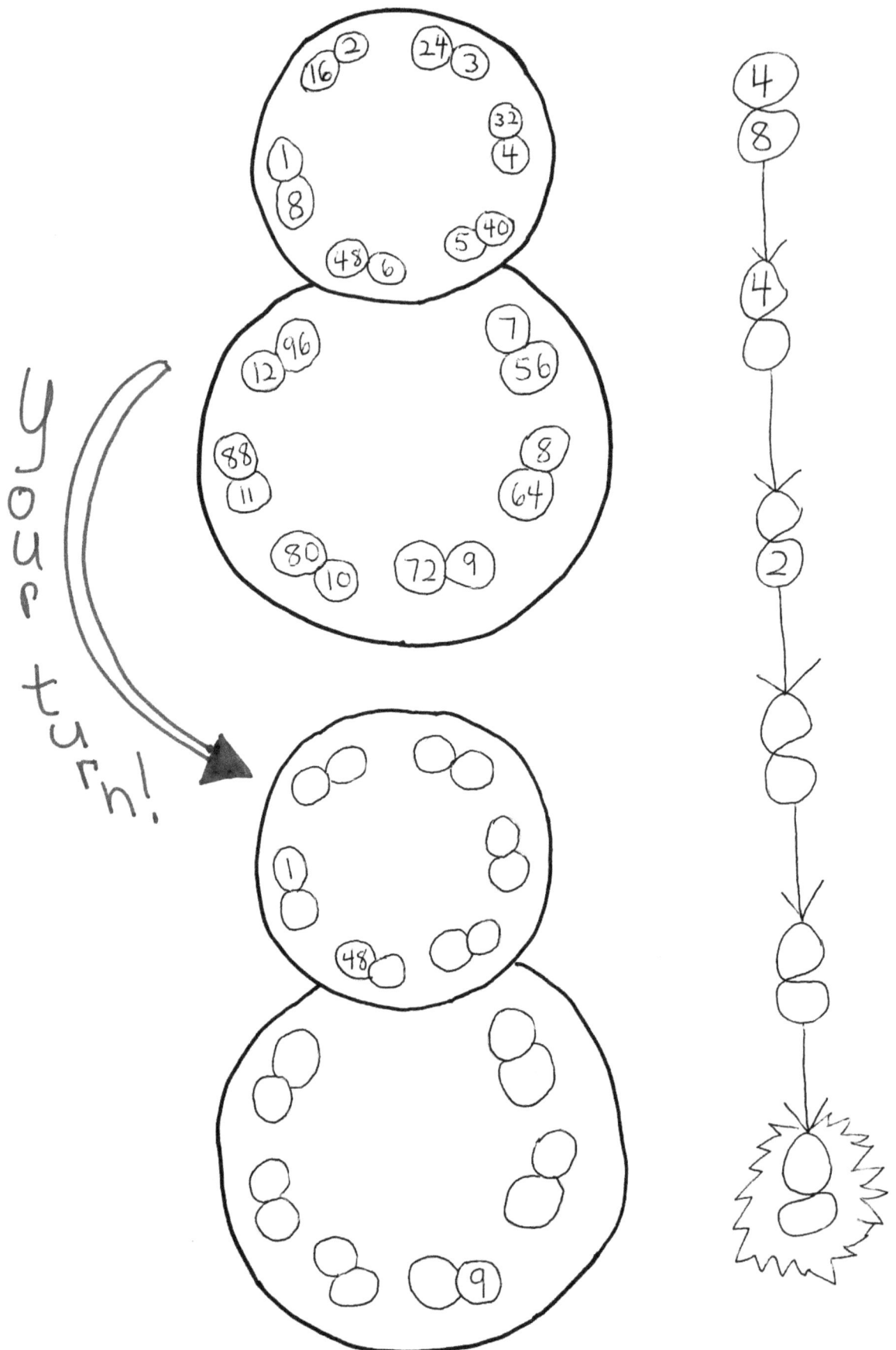

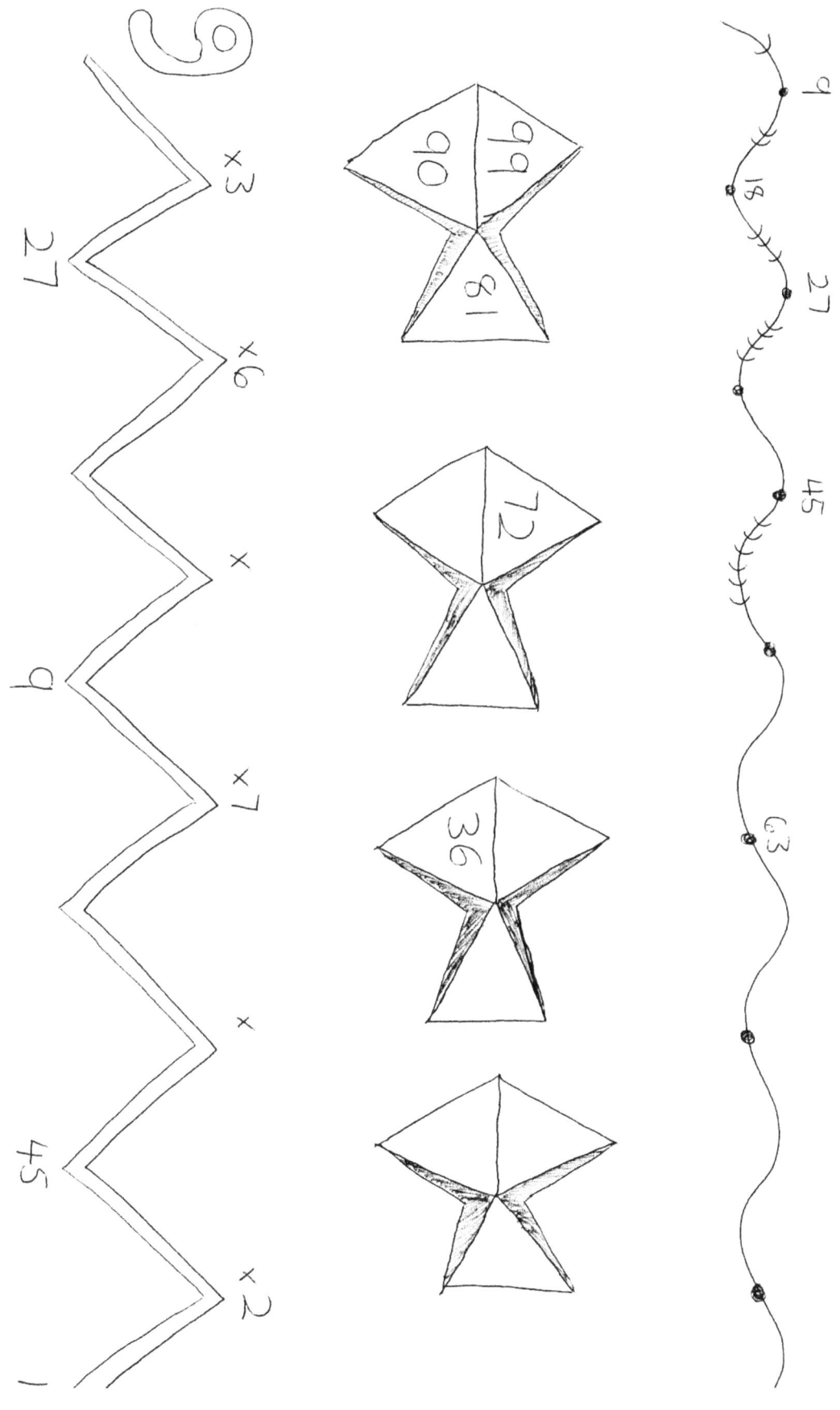

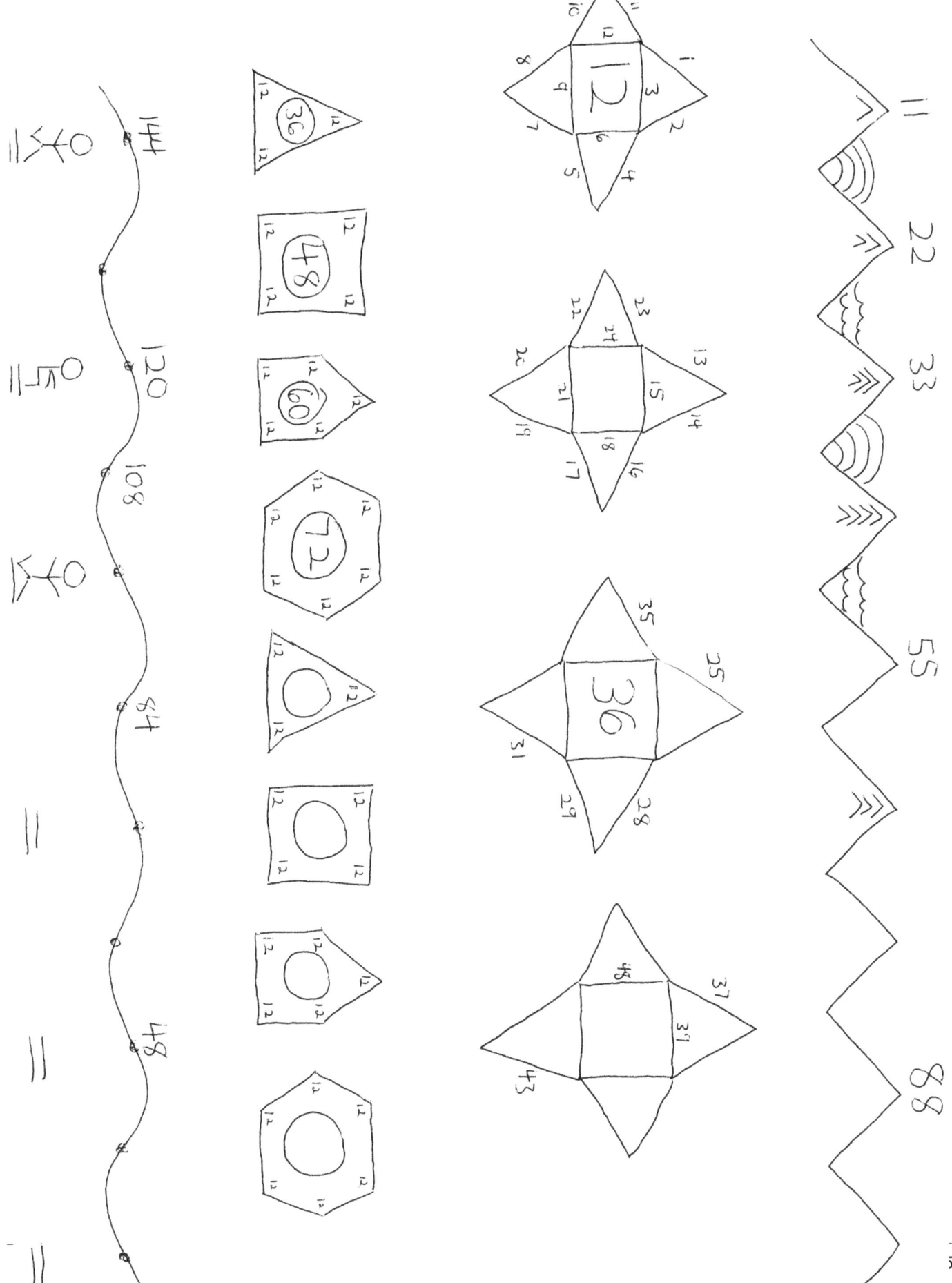

Domino Multiplication

Got dominoes? Then try this simple game.

Players: 2

Materials needed: any set of dominoes

Object of game: Collect as many dominoes as you can.

How to play

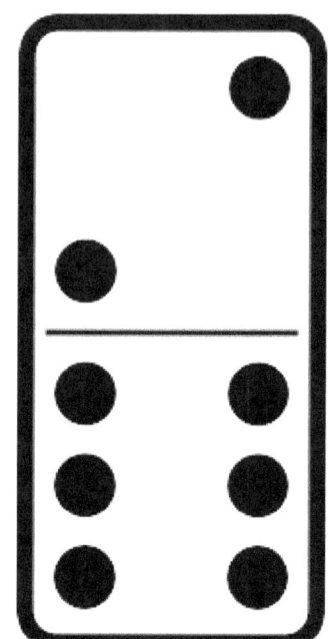

#1: Lay all the dominoes **face down**, either in a stack or a line.

#2: First player chooses a domino.

#3: Player multiplies the two numbers represented on the domino. For example, if it has a two on one side and a five on the other, they do the problem 2 X 5 and say the product, 10.

#4: If they get it right, they get to keep the domino. If not, they have to put it back.

#5: Next player takes a turn.

#6: The winner is the one with the most dominoes at the end.

Hands Up!

To play this game, you just need a friend with two hands who is just learning their multiplication facts, like you.

How to play

#1: Designate a player to call out, "Go!"

#2. Stand facing your partner.

#3: Each of you thinks of a number between one and ten.

#4: The player who is going to say "go" asks if you've got your number.

#5: When you do, the designated player shouts, "Go!", and both of you hold up however many fingers of the number you were thinking of. For example, if you were thinking of the number 6, you hold up six fingers.

#6: Look at your friend's fingers and multiply that number by the number you are holding up. So if you are holding up 6 fingers, and your friend is holding up 4 fingers, you multiply 6 X 4 in your head.

#7: Call out the answer as soon as you have it.

#8: The winner is the person who calls out the most answers first after a certain number of plays.

The End Of The Grid

This is similar to the Race To 100 game, only…different.

Players: 2

Materials needed:

- One Numbers To 100 grid (next page)
- Two game markers, one for each player
- A deck of playing cards (or 1-9 cards that you have made for previous games). Remove face cards.
- Scratch paper

Object of game: Be the first player to get the closest to 100 with both game pieces after ten turns.

How to play

#1: Set the game pieces just outside the top left corner of the table, next to the 1.

#2: Shuffle the cards and set them in a stack between the players, face down.

#3: Player 1 draws the top two cards and multiplies them together. She moves her game piece to the answer.

#4: Player 2 takes a turn doing the same thing. Then, one of the players makes a tally mark on the scratch paper to indicate that one whole turn has been completed by both players.

#5: Player 1 takes her next turn. After she multiplies the numbers on the two cards, she adds that product to the number where her game piece is sitting. For example, if the two cards she drew the first time were a 3 and a 7, her game piece would be on 21. If the second time she drew a 6 and a 3, 6x3=18 so she would add the 18 to the 21. The sum is 39, so she would move her game piece to 39.

#6: If the sum of the two numbers (the number that the game piece is sitting on, plus the product of the two cards) is ever more than 100, the player must stay on the square where he is.

#7: Continue the game until there are 10 tally marks on the paper. The winner is whoever is closest to 100 after each player has had ten turns.

Numbers To 100

1	2	3	4	5	6	7	8	9	10
11	12	13	14	15	16	17	18	19	20
21	22	23	24	25	26	27	28	29	30
31	32	33	34	35	36	37	38	39	40
41	42	43	44	45	46	47	48	49	50
51	52	53	54	55	56	57	58	59	60
61	62	63	64	65	66	67	68	69	70
71	72	73	74	75	76	77	78	79	80
81	82	83	84	85	86	87	88	89	90
91	92	93	94	95	96	97	98	99	100

Multiplication Table Bingo

This game is a combination of Bingo and Odd Man Out.

Players: 2-4

Materials needed:

- A multiplication table for each player
- Something to cover the squares on the table – large dried beans, small buttons, etc.
- That ubiquitous deck of cards (if you use playing cards, remove the jokers and face cards)

What does **ubiquitous** mean? Well, look it up in the dictionary. Then use the word at dinner tonight. Your family will think you've grown a hundred extra brain cells today!

Object of the game: Be the first to cover an entire row or an entire column

How to play

#1: Shuffle the cards and set them in a stack, face down, between the players.

#2: Player 1 draws the top two cards and multiplies the numbers on the cards together.

#3: The player places a marker on the corresponding square on her multiplication table.

#4: Player 2 takes a turn, and so on.

#5: Game continues until a player has covered all the squares in either a row or column on her table.

Puzzle Pictures

On the next five pages, you will see a Multiplication Table. You will also see a bunch of multiplication problems.

Solve those problems to make a picture on the Table! Here's how:

The first number in the problem is one of the numbers going across at the top of the Table. So, put one finger there. The second number in the problem is one of the numbers going down the side. Put another finger there. Move the fingers in a straight line until they meet. Color the square where your fingers meet.

For example, if the problem is 6 X 2, you will end up coloring the square with the 12 that is just under the 6 at the top of the Table.

It is important to do this right, or you will end up coloring the wrong squares!

After you have solved all the problems and colored all of the squares that represent the answers, you will see a message or picture of something.

Ready? Then...

let's go!

Solve the following multiplication problems. Color the squares with the answers and see what appears!

6x2, 5x3, 4x4, 3x5, 3x6, 3x7, 3x8, 3x9, 3x10, 7x3, 8x4, 9x5, 9x6, 9x7, 9x8, 9x9, 9x10, 7x9, 7x10, 5x6, 5x7, 6x6, 6x7

×	1	2	3	4	5	6	7	8	9	10	11	12
1	1	2	3	4	5	6	7	8	9	10	11	12
2	2	4	6	8	10	12	14	16	18	20	22	24
3	3	6	9	12	15	18	21	24	27	30	33	36
4	4	8	12	16	20	24	28	32	36	40	44	48
5	5	10	15	20	25	30	35	40	45	50	55	60
6	6	12	18	24	30	36	42	48	54	60	66	72
7	7	14	21	28	35	42	49	56	63	70	77	84
8	8	16	24	32	40	48	56	64	72	80	88	96
9	9	18	27	36	45	54	63	72	81	90	99	108
10	10	20	30	40	50	60	70	80	90	100	110	120
11	11	22	33	44	55	66	77	88	99	110	121	132
12	12	24	36	48	60	72	84	96	108	120	132	144

Solve the following multiplication problems. Color the squares with the answers and see what appears!

2x4, 2x5, 2x6, 2x7, 2x8, 3x6, 4x4, 4x5, 4x6, 4x7, 4x8, 6x4, 6x8, 7x4, 7x5, 7x6, 7x8, 8x8

×	1	2	3	4	5	6	7	8	9	10	11	12
1	1	2	3	4	5	6	7	8	9	10	11	12
2	2	4	6	8	10	12	14	16	18	20	22	24
3	3	6	9	12	15	18	21	24	27	30	33	36
4	4	8	12	16	20	24	28	32	36	40	44	48
5	5	10	15	20	25	30	35	40	45	50	55	60
6	6	12	18	24	30	36	42	48	54	60	66	72
7	7	14	21	28	35	42	49	56	63	70	77	84
8	8	16	24	32	40	48	56	64	72	80	88	96
9	9	18	27	36	45	54	63	72	81	90	99	108
10	10	20	30	40	50	60	70	80	90	100	110	120
11	11	22	33	44	55	66	77	88	99	110	121	132
12	12	24	36	48	60	72	84	96	108	120	132	144

Solve the following multiplication problems. Color the squares with the answers and see what appears!

2x2, 2x3, 2x4, 2x5, 2x6, 3x2, 4x2, 5x4, 5x5, 5x6, 6x6, 7x4, 7x5, 7x6, 9x4, 9x5, 9x6, 10x4, 11x4, 11x5, 11x6

×	1	2	3	4	5	6	7	8	9	10	11	12
1	1	2	3	4	5	6	7	8	9	10	11	12
2	2	4	6	8	10	12	14	16	18	20	22	24
3	3	6	9	12	15	18	21	24	27	30	33	36
4	4	8	12	16	20	24	28	32	36	40	44	48
5	5	10	15	20	25	30	35	40	45	50	55	60
6	6	12	18	24	30	36	42	48	54	60	66	72
7	7	14	21	28	35	42	49	56	63	70	77	84
8	8	16	24	32	40	48	56	64	72	80	88	96
9	9	18	27	36	45	54	63	72	81	90	99	108
10	10	20	30	40	50	60	70	80	90	100	110	120
11	11	22	33	44	55	66	77	88	99	110	121	132
12	12	24	36	48	60	72	84	96	108	120	132	144

Solve the following multiplication problems. Color the squares with the answers and see what appears!

4x5, 3x7, 4x8, 5x9, 6x9, 7x5, 7x8, 8x7

×	1	2	3	4	5	6	7	8	9	10	11	12
1	1	2	3	4	5	6	7	8	9	10	11	12
2	2	4	6	8	10	12	14	16	18	20	22	24
3	3	6	9	12	15	18	21	24	27	30	33	36
4	4	8	12	16	20	24	28	32	36	40	44	48
5	5	10	15	20	25	30	35	40	45	50	55	60
6	6	12	18	24	30	36	42	48	54	60	66	72
7	7	14	21	28	35	42	49	56	63	70	77	84
8	8	16	24	32	40	48	56	64	72	80	88	96
9	9	18	27	36	45	54	63	72	81	90	99	108
10	10	20	30	40	50	60	70	80	90	100	110	120
11	11	22	33	44	55	66	77	88	99	110	121	132
12	12	24	36	48	60	72	84	96	108	120	132	144

Solve the following multiplication problems. Color the squares with the answers and see what appears!

4x2, 4x3, 4x4, 3x4, 2x4, 2x5, 2x6, 3x6, 4x6, 4x7, 4x8, 5x8, 6x8, 6x7, 6x6, 7x6, 8x6, 8x5, 8x4, 7x4, 6x4, 6x3, 6x2, 5x2

×	1	2	3	4	5	6	7	8	9	10	11	12
1	1	2	3	4	5	6	7	8	9	10	11	12
2	2	4	6	8	10	12	14	16	18	20	22	24
3	3	6	9	12	15	18	21	24	27	30	33	36
4	4	8	12	16	20	24	28	32	36	40	44	48
5	5	10	15	20	25	30	35	40	45	50	55	60
6	6	12	18	24	30	36	42	48	54	60	66	72
7	7	14	21	28	35	42	49	56	63	70	77	84
8	8	16	24	32	40	48	56	64	72	80	88	96
9	9	18	27	36	45	54	63	72	81	90	99	108
10	10	20	30	40	50	60	70	80	90	100	110	120
11	11	22	33	44	55	66	77	88	99	110	121	132
12	12	24	36	48	60	72	84	96	108	120	132	144

Your turn!

Now it's YOUR turn to create a picture puzzle for someone else to solve! Below is yet another multiplication table. With a pencil, **very** lightly outline the squares that will form a short word or a simple picture.

Then, figure out the multiplication fact that goes with each square. Remember that the first factor in your problem will be the number going along the top of the table. Write the facts down below the table.

When you are sure you have all the facts that somebody needs to solve the puzzle, erase the pencil marks you made earlier and pass the puzzle on. Have fun!

×	1	2	3	4	5	6	7	8	9	10	11	12
1	1	2	3	4	5	6	7	8	9	10	11	12
2	2	4	6	8	10	12	14	16	18	20	22	24
3	3	6	9	12	15	18	21	24	27	30	33	36
4	4	8	12	16	20	24	28	32	36	40	44	48
5	5	10	15	20	25	30	35	40	45	50	55	60
6	6	12	18	24	30	36	42	48	54	60	66	72
7	7	14	21	28	35	42	49	56	63	70	77	84
8	8	16	24	32	40	48	56	64	72	80	88	96
9	9	18	27	36	45	54	63	72	81	90	99	108
10	10	20	30	40	50	60	70	80	90	100	110	120
11	11	22	33	44	55	66	77	88	99	110	121	132
12	12	24	36	48	60	72	84	96	108	120	132	144

Multiplying Numbers Containing Zeros At The End

Starting with the digit 1...

Hey! How would you like to be known as the Smartest Kid On The Block? How would you like to be able to multiply **GIGANTIC** numbers...*in your head?*

Yes? Okay! Let me show you yet another fun, monumentally cool multiplication trick.

But first, tell me the rule for multiplying by 10. If you multiply any number by ten, the product (answer) will be_____.

Well, guess what? The same rule applies for multiplying by 100, 1,000, 10,000 and even **BIGGER** numbers that start with the digit one and are followed by only zeros.

I'm going to start a couple of patterns below. The column on the left is one pattern, and the one on the right is a different pattern. Fill in the missing numbers.

1 X 100 = 100					1 X 1,000 = 1,000

2 X 100 = 200					2 X 1,000 = 2,000

3 X 100 = 300					3 X 1,000 = _____

4 X 100 = _____					4 X 1,000 = _____

5 X 100 = _____					5 X 1,000 = _____

6 X 100 = _____					___X 1,000 = 6,000

___X 100 = 700					___X 1,000 = 7,000

___X 100 = 800					8 X 1,000 = _____

36 X 100 = 3600					42 X 1,000 = 42,000

55 X 100 = _____				65 X 1,000 = _____

76 X 100 = _____				97 X 1,000 = _____

109

Another super-cool trick! When you multiply any given number by a number that starts with a one and then has only zeros after it, the product is...what? Finish that statement!

Now, go teach that rule to your little brother or sister or the poodle next door so that you will **really** learn it.

But come right back, because then I have another exciting trick to show you!

Starting with a digit other than 1...

Now you know how to multiply any number by 10, 100, 1,000, and so on. What if you wanted to multiply a number by 20? Or 300? 4,000? And what if you wanted to do it **in your head**, no calculator, no paper and pencil?

No problem! It's basically the same rule. I'm going to start three patterns below. Examine them carefully, and you will be able to fill in the missing numbers.

20 X 2 = 40	200 X 3 = 600	2,000 X 2 = 4,000
20 X 20 = 400	300 X 3 = 900	3,000 X 2 = 6,000
20 X 3 = 60	400 X 3 = _____	4,000 X 2 = _____
20 X 30 = 600	500 X 3 = _____	5,000 X 2 = 10,000
20 X 4 = _____	600 X 3 = _____	6,000 X 2 = _____
20 X 40 = _____	700 X 3 = _____	7,000 X 2 = _____
20 X 5 = _____	800 X 3 = _____	8,000 X 2 = _____
20 X 50 = _____	900 X 3 = _____	9,000 X 2 = _____

Did you get the rule for multiplying numbers that have zeros tacked onto the end? This *one* time I'll help you out, in case you didn't: Ignore the zeros and multiply the other digits. Then, however many zeroes are at the end of both numbers, stick those on to the end, and you have your product!

Let's do 2,000 X 4. 2 X 4 = 8. And the 2,000 has three zeros after the 2. So you stick three zeros after the 8, giving you the answer of 8,000.

What about 30 X 30? Well, 3 X 3 = 9. Each number has one zero tacked on the end of it, for a total of two zeros. So you stick two zeros after the 9 to make 900. 30 X 30 = 900.

Ah, now the light bulb is coming on! Go back and finish the above problems if you need to. Then let's switch our operation gears for a minute, from multiplication to division.

What Is Division?

By now, you know what multiplication is. You should be close to mastering the multiplication facts through 12. You even know several tricks and fun patterns related to those facts!

So now, you are ready to

divide and conquer!

Okay, so, just divide. You are going to learn the operation of *division*. What is that?

First, think about the word "divide". What does that mean? Now, think about what multiplication is. Multiplication is repeated_____.

Wonderful! I'm so glad you've been paying attention. So, just like subtraction is the opposite of addition, division is the opposite of multiplication. So if multiplication is repeated **addition**, what is division? Division is repeated_____.

****

Did you say "repeated subtraction"? Fantabulous! Pat yourself on the back!

Let's look at a simple division problem: 6 ÷ 2 = 3. You read that as "Six divided by two equals three." What that means is that you have six things, and you divide them into two equal groups. When you do that, you end up with three things in each of those two groups.

Here's how it looks as a repeated subtraction sentence: 6 - 2 - 2 - 2 = 0.

Division is subtracting the same number over and over until the answer is zero. How many times did you have to subtract two in that above example in order to reach zero?

Exactly! Three times. So, 6 ÷ 2 = 3. _{Sometimes you will see this symbol: / It also means "divided by, just like this symbol: ÷}

By the way, the big number you are dividing into parts is called the <u>divisor</u>, while the smaller number is the <u>dividend</u>. The answer of a division problem is called the <u>quotient</u>.

Here's what the division sentence, 6 ÷ 2 = 3, looks like in picture form:

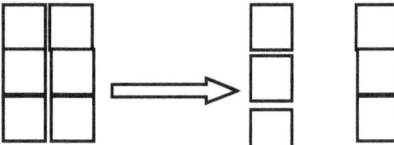

When you have six squares, and divide them into two equal groups, you end up with three squares in each group.

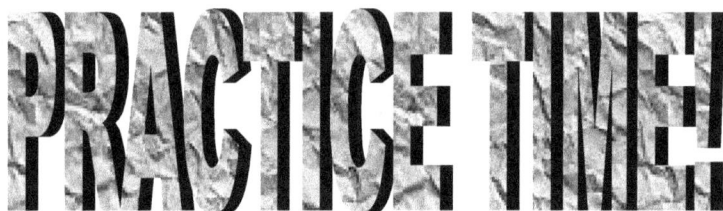

Now you're going to practice the concept of division just like you practiced multiplication back at the beginning of the book. Take out the number bars you made and get ready to do some dividing!

But remember – division is the opposite of multiplication. With multiplication, you start with one bar and add a given number of those bars to make a rectangle, and then count to see how many squares you have when you're finished. With division, you start with a rectangle, then break it up.

Look back at the example above, 6 ÷ 2 = 3. You start with a rectangle that has six total squares. Then you break apart that rectangle into two groups. Finally, you count how many squares end up in each group. It's a little trickier than multiplication, because you might choose number bars that don't break up exactly the way you planned them to. If so, grab some wooden blocks or cubes out of your room, or simply draw pictures of objects that you can divide into groups.

When you have all your materials ready, go ahead and do the following division problems.

15 ÷ 5   21 ÷ 7   18 ÷ 9   8 ÷ 2   9 ÷ 3   24 ÷ 6

Want to know an even *easier* way to figure out the answers to those division problems? Use the Multiplication Table! Here's how: find the divisor inside the table (in 6 ÷ 2, six is the divisor). You may find it more than once. Which one do you choose? The one that, if you move your finger either straight up or straight back to the left, you run into the dividend (in 6 ÷ 2, two is the dividend). So if you're trying to figure out 24 ÷ 6, find the 24 on the table that is either straight across from or straight down from 6. Then, move your finger straight up or across to find the number that you multiply by 6 to get 24. You should hit 4. That is the *quotient*, or the answer to 24 ÷ 6.

I **TOLD** you it was easy! Now, go back to those division problems you worked on with the manipulatives (number bars or blocks or whatever), and do them again using the Multiplication Table.

As soon as you're done, come back to meet some new families!

Multiplication and Division Fact Families

Chances are high that back when you learned about addition and subtraction, somebody taught you about fact families. A fact family is a group of two or four number sentences that all have the same three numbers in them.

For example, here's the fact family for 3 + 4:

$$3 + 4 = 7, \quad 4 + 3 = 7, \quad 7 - 3 = 4, \quad 7 - 4 = 3$$

All of those number sentences contain the numbers 3, 4 and 7, just in a different order. So they are all related. They are **families**.

You can make fact families using multiplication and division, too. For the next several pages, you will find all those fact families grouped together in boxes. But not all the number sentences will be complete! Your mission is to fill in the missing numbers in each fact family.

Ready? Got your helmet and your sword? I mean, your gun and your bulletproof vest? No, no, you just need a pencil and your brain in order to succeed with your mission! If you are a right-brained, picture-thinking student, use the number bars as you go, too.

2 X 3 = 6	3 X ____ = 6
6 ÷ 3 = ____	____ ÷ 2 = 3

2 X 4 = 8	____ X 2 = 8
8 ÷ ____ = 4	8 ÷ 4 = ____

2 X 5 = 10   5 X ____ = 10

____ ÷ 2 = 5   10 ÷ ____ = 2

---

2 X 6 = 12   6 X 2 = 12

12 ÷ 2 = ____   12 ÷ ____ = 6

---

2 X 7 = 14   7 X 2 = ____

____ ÷ 7 = 2   14 ÷ ____ = 7

---

2 X 8 = 16   8 X 2 = 16

16 ÷ 8 = ____   16 ÷ ____ = 2

---

2 X 9 = 18   ____ X 2 = 18

____ ÷ 9 = 2   18 ÷ 2 = ____

---

2 X 10 = 20   10 X ____ = 20

20 ÷ 10 = ____   ____ ÷ 2 = 10

2 X 11 = 22	11 X 2 = _____
22 ÷ ____ = 2	22 ÷ 11 = _____

2 X 12 = 24	____ X 2 = 24
____ ÷ 12 = 2	24 ÷ ____ = 12

3 X 3 = 9	9 ÷ ____ = 3

3 X 4 = 12	____ X 3 = 12
12 ÷ ____ = 4	12 ÷ 4 = _____

3 X 5 = 15	5 X _____ = 15
15 ÷ 3 = _____	_____ ÷ 5 = 3

3 X 6 = 18	6 X _____ = 18
____ ÷ 3 = 6	18 ÷ 6 = _____

3 X 7 = 21	7 X 3 = _____
21 ÷ ____ = 7	21 ÷ 7 = _____

3 X 8 = 24   8 X _____ = 24

24 ÷ 3 = _____   24 ÷ _____ = 8

---

3 X 9 = 27   9 X 3 = _____

_____ ÷ 3 = 9   27 ÷ 9 = _____

---

3 X 10 = 30   10 X _____ = 30

30 ÷ _____ = 10   _____ ÷ 10 = 3

---

3 X 11 = 33   11 X 3 = 33

33 ÷ 3 = _____   33 ÷ 11 = _____

---

3 X 12 = 36   _____ X 3 = 36

_____ ÷ 3 = 12   36 ÷ _____ = 3

---

4 X 4 = 16   _____ ÷ 4 = 4

4 X 5 = 20	5 X _____ = 20
_____ ÷ 4 = 5	20 ÷ 5 = _____

4 X 6 = 24	_____ X 4 = 24
24 ÷ _____ = 6	24 ÷ 6 = _____

4 X 7 = 28	7 X _____ = 28
_____ ÷ 4 = 7	28 ÷ _____ = 4

4 X 8 = 32	_____ X 4 = 32
32 ÷ _____ = 8	32 ÷ 8 = _____

4 X 9 = 36	9 X 4 = _____
_____ ÷ 4 = 9	36 ÷ _____ = 4

4 X 10 = 40	10 X _____ = 40
40 ÷ _____ = 10	40 ÷ 10 = _____

$4 \times 11 = 44$ _____ $\times 4 = 44$

_____ $\div 4 = 11$ $44 \div 11 =$ _____

$4 \times 12 = 48$ $12 \times 4 =$ _____

$48 \div$ _____ $= 12$ $48 \div$ _____ $= 4$

$5 \times 5 = 25$ $25 \div$ _____ $= 5$

$5 \times 6 = 30$ _____ $\times 5 = 30$

$30 \div 5 =$ _____   _____ $\div 6 = 5$

$5 \times 7 = 35$ $7 \times$ _____ $= 35$

_____ $\div 5 = 7$ $35 \div 7 =$ _____

$5 \times 8 = 40$ _____ $\times 5 = 40$

$40 \div 5 =$ _____ $40 \div$ _____ $= 5$

$5 \times 9 = 45$ $9 \times 5 =$ _____

_____ $\div 5 = 9$ $45 \div 9 =$ _____

5 X 10 = 50   10 X _____ = 50

50 ÷ 5 = _____   50 ÷ 10 = _____

---

5 X 11 = 55   _____ X 5 = 55

_____ ÷ 5 = 11   55 ÷ 11 = _____

---

5 X 12 = 60   12 X 5 = _____

60 ÷ 5 = _____   60 ÷ _____ = 5

---

6 X 6 = 36   36 ÷ 6 = _____

---

6 X 7 = 42   7 X 6 = _____

42 ÷ _____ = 7   _____ ÷ 7 = 6

---

6 X 8 = 48   _____ X 6 = 48

_____ ÷ 6 = 8   48 ÷ 8 = _____

6 X 9 = 54   9 X _____ = 54

54 ÷ 6 = _____   54 ÷ 9 = _____

---

6 X 10 = 60   _____ X 6 = 60

_____ ÷ 6 = 10   60 ÷ 10 = _____

---

6 X 11 = 66   11 X _____ = 66

66 ÷ 6 = _____   66 ÷ 6 = _____

---

6 X 12 = 72   _____ X 6 = 72

_____ ÷ 6 = 12   72 ÷ _____ = 6

---

7 X 7 = 49   _____ ÷ 7 = 7

---

7 X 8 = 56   8 X 7 = _____

_____ ÷ 7 = 8   56 ÷ _____ = 7

---

7 X 9 = 63   _____ X 7 = 63

63 ÷ 7 = _____   _____ ÷ 9 = 7

7 X 10 = 70	10 X _____ = 70
_____ ÷ 7 = 10	70 ÷ 10 = _____

7 X 11 = 77	_____ X 7 = 77
77 ÷ _____ = 11	77 ÷ _____ = 7

7 X 12 = 84	12 X 7 = _____
_____ ÷ 7 = 12	84 ÷ _____ = 7

8 X 8 = 64	64 ÷ 8 = _____

8 X 9 = 72	_____ X 8 = 72
72 ÷ 8 = _____	72 ÷ _____ = 8

8 X 10 = 80	10 X _____ = 80
_____ ÷ 8 = 10	80 ÷ 10 = _____

8 X 11 = 88	_____ X 8 = 88
88 ÷ _____ = 11	88 ÷ _____ = 8

8 X 12 = 96	12 X 8 = _____
_____ ÷ 8 = 12	96 ÷ _____ = 8

9 X 9 = 81	_____ ÷ 9 = 9

9 X 10 = 90	_____ X 9 = 90
90 ÷ _____ = 10	90 ÷ 10 = _____

9 X 11 = 99	11 X _____ = 99
_____ ÷ 9 = 11	99 ÷ _____ = 9

9 X 12 = 108	_____ X 9 = 108
108 ÷ 9 = _____	108 ÷ 12 = _____

10 X 10 = 100	100 ÷ _____ = 10

10 X 11 = 110   ____ X 10 = 110

110 ÷ 10 = ____   ____ ÷ 11 = 10

---

10 X 12 = 120   12 X 10 = ____

____ ÷ 10 = 12   120 ÷ 12 = ____

---

11 X 11 = 121   121 ÷ ____ = 11

---

11 X 12 = 132   ____ X 11 = 132

132 ÷ ____ = 12   ____ ÷ 12 = 11

---

12 X 12 = 144   144 ÷ ____ = 12

The Answer Is Always 11

Back when you were learning the 11X facts, I taught you a neat pattern that happens when you multiply 11 by other numbers whose digits consist only of 1's. Remember, 11 X 111 and so on.

Well, now that you understand division, I've got another fun trick to do with the number 11. Once again, you need a calculator.

#1. Think of a 3-digit number. (146, for example.)

#2. Type that number into the calculator twice. (Using our example, you would type in 146146.)

#3. Divide that number by 7. (146146 ÷ 7 = 20,878)

#4. Divide the answer you get in step #3 by 13. (20,878 ÷ 13 = 1606)

#5. Finally, divide the answer you get in step #4 by the original 3-digit number. (1606 ÷ 146 = 11)

If you did it correctly, the answer will be 11 every single time! **No matter which 3-digit number you choose!** Totally cool, huh? Now, go try that trick on a grown-up (but be nice and let them use your calculator).

Odd Man Out

Players: 2

Materials needed:

- The usual cards with 1-9
- Numbers To 100 Grid (next page)
- A blue crayon and a red crayon (or colored pencils)

Object of game: Be the person who fills in the last number in a column or row of numbers.

How to play

#1: Shuffle the cards. Place them face down between the players.

#2: Player 1 draws the top two cards and multiplies the numbers on the cards together.

#3: If the product of those numbers turns out to be an even number, he colors the square on the grid with that number blue. If it is an odd number, he colors the square red. For example, let's say he draws a card with a 5 and a card with a 6. The product of 5 and 6 is 30. Thirty is an even number, so he would color the 30 on the number grid blue.

#4: Player 2 takes a turn.

#5: If a square is already colored, the next player takes her turn as usual.

#6: After the first two turns, if a player wants, he can do a division problem instead of multiplication. In that case, he will draw only one card instead of two. Then he will make a division fact out of the last square colored. So if the last square colored was 15, he might choose to draw one card. If he draws a 3, he would do the problem 15÷3, which is 5, and so color the five.

If a player draws a number that does not divide exactly into the last square colored, he can color the number that would be closest. For example, if the last square colored is 15, and he draws a card with a 2 on it, the division problem would be 15÷2. You can't divide 15 exactly into 2, but you can divide both 14 and 16 into 2. 14÷2=7 and 16÷2=8, so he could choose to color either the 7 square OR the 8 square.

A player might choose to do this if it might give him a number which would be one of the last ones in a column or row to color.

#7: This game could take a long time to end, so feel free to change the rules about when there is a winner. For example, you might say that instead of having to color in an entire row or column, you only have to color five numbers in a row.

Numbers To 100

1	2	3	4	5	6	7	8	9	10
11	12	13	14	15	16	17	18	19	20
21	22	23	24	25	26	27	28	29	30
31	32	33	34	35	36	37	38	39	40
41	42	43	44	45	46	47	48	49	50
51	52	53	54	55	56	57	58	59	60
61	62	63	64	65	66	67	68	69	70
71	72	73	74	75	76	77	78	79	80
81	82	83	84	85	86	87	88	89	90
91	92	93	94	95	96	97	98	99	100

Remainders

QUICK! What's the answer to this problem:

$$15 \div 2 = ?$$

Because division is the opposite of multiplication, if you don't know a basic division fact you can always try turning it into a multiplication fact to figure out the answer. In this case, you would ask yourself, "Two times what equals fifteen?"

The problem is, you would come up empty! **Nothing** times two equals fifteen. If you have fifteen apples, you can't make groups of apples that each have exactly the same number. That's where remainders come in. Leftovers. Many division problems end up with them. Like $15 \div 2$.

Look at this picture:

+ + + + + + + + + + + + + + +

There are fifteen plus signs. Now, try to divide those plus signs into two equal groups.

+ + + + + + +   + + + + + + +

But those groups aren't equal, are they? There are eight plus signs in the first group, and seven in the second. To divide is to make groups that have equal numbers in them. So you have to make sure the groups have equal numbers.

How do you do that? Make the groups as equal as you can. Start by thinking of a division problem that **does** turn out equal groups that is very close to the problem you are working on. But the divisor (first number in the problem) must be **smaller** than the divisor in the original problem. In this example, it must be smaller than 15.

Start at the number that comes just before 15, and, if necessary, work your way down. In this case, 14 is the number that comes just before 15.

Can you divide 2 into 14 exactly? It turns out that you can. $14 \div 2 = 7$.

If we divide fourteen plus signs into two equal groups, each group will have seven plus signs in it.

+ + + + + + +   + + + + + + +

But 15 ÷ 2 does **not** come out even. What do we end up with, then?

+ + + + + +   + + + + + +   +

We still need to end up with two equal groups. In the above picture, we have two equal groups. Each group has seven plus signs. But because you cannot divide 15 into two exactly equal groups, there is one extra plus sign. That is the **remainder**. When you divide two into fifteen, you end up with two groups of seven, with a remainder of one.

Here's how we write that: 15 ÷ 2 = 7 R 1 (the "R" means "remainder", or leftover).

Let me give you another example: 11 ÷ 4.

+ + + + + + + + + + +

Can you divide the above plus signs into four **equal** groups? No? So what's your next step?

Try 10 ÷ 4. Will that turn out equal groups? No, so then try 9 ÷ 4. No again. Keep moving down. How about 8 ÷ 4?

Yes, if you do *that* problem, you will end up with four equal groups.

+ +   + +   + +   + +

8 ÷ 4 = 2, because when you divide eight plus signs into four equal groups, you end up with two plus signs in each group.

11 ÷ 4 is also going to equal 2, but have a remainder.

+ + + + + + + + + + +

Use a pencil to divide the above plus signs into four equal groups. You should end up with two in each group. How many plus signs are leftover?

+ +   + +   + +   + +   + + +

You end up with two plus signs in each group, with three plus signs leftover. So three is the remainder.

11 ÷ 4 = 2 R 3

I'm going to have you practice a few division problems with remainders on your own. But first...SHHH! I have a secret to tell you. **This is very important – please listen carefully!**

The remainder should <u>ALWAYS</u> be smaller than the dividend. In the example above, 11 ÷ 4, 4 is the dividend (11 is the divisor).

If you end up with a remainder that is BIGGER than the dividend, you need to try again!

Got it? All right! So, let's try a few more division with remainder problems. Feel free to use beans, cubes, buttons, or whatever to figure them out.

7 ÷ 2 = _____ R ___   16 ÷ 5 = _____ R ___

10 ÷ 3 = _____ R ___   21 ÷ 6 = _____ R ___

37 ÷ 7 = _____ R ___   46 ÷ 8 = _____ R ___

Great job! That was a lot of hard work, but you're getting it, so give yourself a pat on the back.

Coming up next is a fun game to practice what you just learned – dividing, with and without remainders. So take a nice brain break and then come back ready for some exciting action!

Dice Division

Players: 2

Materials needed:

- Two dice
- Multiplication table, for reference
- Piece of paper

Object of game: To have the most tally marks after each player has taken ten turns.

How to play

#1: Both players write their names at the top of the paper, on opposite sides.

#2: Player 1 throws both dice and puts the two numbers together to make a two-digit number. For example, if he rolls a 4 and a 2, he can make either 42 or 24.

#3: Player must decide if he can make a division problem out of the two-digit number. He may look on the multiplication table to determine this. If he can, he says the division problem out loud ("Forty-two divided by seven equals six). Then he writes a tally mark under his name on the paper.

If he cannot make a division problem, Player 2 takes a turn and Player 1 does not earn a tally mark.

#4: Also make a tally mark at the bottom of the page to keep track when both players have each had a turn.

#5: Keep playing until they have had ten turns each. Winner is whoever has the most tally marks (for being able to make a division problem) under his name.

A Superstar Multiplication Trick

Okay, by now you should have learned most of the basic multiplication (even basic division) facts. If not, keep working on them, because this trick won't work unless you've got the basic facts **super-glued** into your brain.

I'm going to teach you how to multiply a two-digit number by a one-digit number...in your head. And when you learn that, you will be a

I'll start off easy. 16 X 3 = ?

What you do is break each number down into its place value. Like this:

$$\begin{array}{rl} 16 & = 10 + 6 \\ \times\ 3 & \ \times\ \ 3 \end{array}$$

Then, multiply the units place of the two-digit number by the one-digit number:

6 X 3 = 18

Next, multiply the tens place of the two-digit number by the one-digit number:

10 X 3 = 30

Finally, add those two products, 18 and 30, together:

```
  18
 +30        SO, 16 X 3 = 48!
  48
```

Know what makes this easy? The rule for multiplying by numbers with zeros! You end up with at least one number that has a zero in the units place, which makes it really easy to add the numbers in your head!

Let's try another example:

$$25 \times 4 = ?$$

Break it down: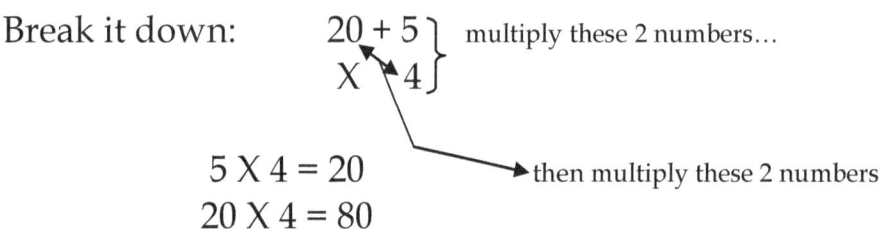

$5 \times 4 = 20$
$20 \times 4 = 80$

Add up the two products:

```
  20
 +80
 100
```

It will probably take you some practice before you can solve problems like these in your head, so, hey, let's do a few more for practice!

I'll leave some space after each problem for you to write down whatever you need to write down to help. But remember – this is **mental** math, so try to do as much of it in your head as you can.

26 X 5

74 X 2

43 X 4

52 X 7

33 X 8

Fantastic! Are you getting faster? You will, as you practice more. When you're ready, ask a friend or parent to give you a couple of problems like this. Solve them without pencil and paper, and they will know that you are a

Multiplying 2-Digit By 1-Digit Numbers On Paper

Once you can multiply problems like 54 X 9 in your head, why do you need to learn to do them on paper?

Because you might have a **MEAN** teacher who will one day give you even bigger problems, and you will have to show your work on paper.

Seriously. Even if you can do 1,287 X 342 in your head in ten seconds, some teachers won't count a correct answer unless you **show your work.**

Bummer, huh? But I'm going to teach you how to show your work, even while encouraging you to work on learning to multiply large numbers in your head. And since the rules are the same no matter how big the numbers are, it makes sense to start small, right?

In fact, let's not just start small. Let's start *easy*. On the next page is the multiplication problem we are going to start with, and how to solve this problem using base ten blocks. (Remember those? The little units cubes, the long ten rods, the hundreds squares?)

After that page I am going to explain how to solve multiplication problems on paper. Please go back to the page with the base ten blocks as we do each step on paper, so that you will really, really understand how it works.

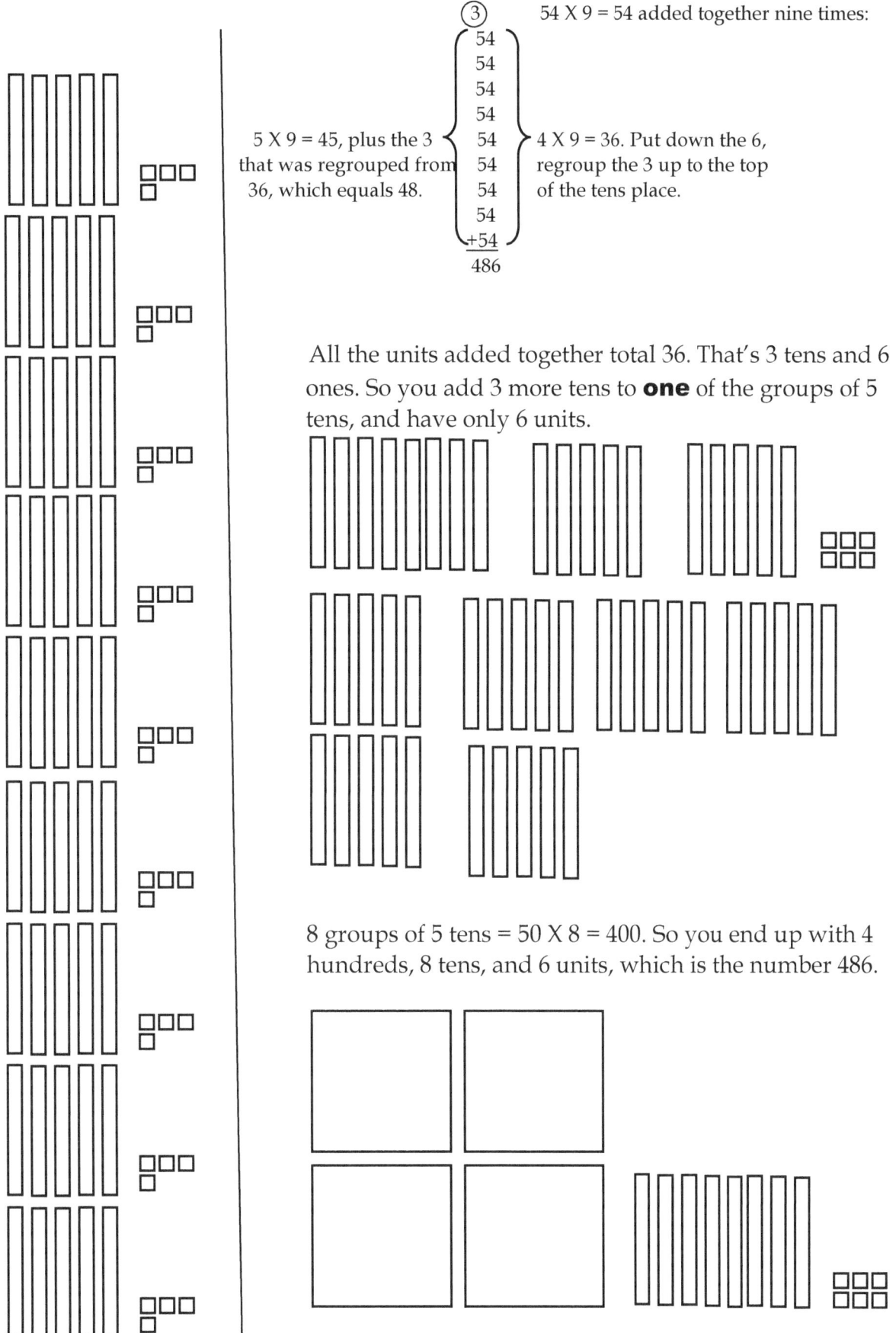

$5 \times 9 = 45$, plus the 3 that was regrouped from 36, which equals 48.

$54 \times 9 = 54$ added together nine times:

$4 \times 9 = 36$. Put down the 6, regroup the 3 up to the top of the tens place.

All the units added together total 36. That's 3 tens and 6 ones. So you add 3 more tens to **one** of the groups of 5 tens, and have only 6 units.

8 groups of 5 tens = $50 \times 8 = 400$. So you end up with 4 hundreds, 8 tens, and 6 units, which is the number 486.

137

That's what 54 X 9 looks like in picture form. On paper, multiplying numbers with more than one digit is basically the same as addition with regrouping. Start with the units place, then move over and multiply the tens place, and so on.

What doesn't fit in the units place needs to be regrouped (or carried) over to the tens place, and what doesn't fit in the tens place needs to be regrouped over to the hundreds place, and so on.

4 X 9 = 36. Put the 6 down in the units place of the answer, then regroup the 3 tens up to the tens place.

```
  3
 54   36
X 9
  6
```

Now, solve 5 X 9, which equals 45.

STOP RIGHT THERE! Before you write down 45 next to the 6, remember the three that you regrouped to the tens place!

Here's where a lot of kids get confused. They think they need to <u>multiply</u>, and so do 45 X 3.

But it's much easier than that! Whenever you regroup a digit in a multiplication problem, you simply have to **add** it to the place you just multiplied.

In this problem, you just multiplied 5 X 9, which is 45, right? But you also regrouped 3 tens from the number 36 when you multiplies 4 X 9. Just add that 3 to the 45.

```
  3
 54
X 9   ← 5 X 9 = 45, 45 + 3 = 48
486
```

Let's do one more example, and then it will be time for a fun game!

First, see if you can work out the product to this problem the way I showed you on the previous page, before you read the explanations on how to solve it.

$$\begin{array}{r} 47 \\ \times 5 \\ \hline \end{array}$$

Start with the units place. 7 X 5 = 35. Put down the 5, and regroup the 3 to the tens place, above the 4 in 47.

$$\begin{array}{r} 3 \\ 47 \quad 35 \\ \times 5 \\ \hline 5 \end{array}$$

Now, multiply the tens place of the first number by 5:

4 X 5 = 20, then add the 3 that you regrouped from the number 35.

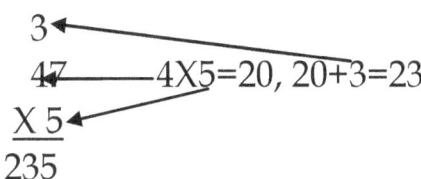

$$\begin{array}{r} 3 \\ 47 \quad \text{—4X5=20, 20+3=23} \\ \times 5 \\ \hline 235 \end{array}$$

Are you getting it? Awesome! Let's get some practice now with a game!

Solve It!

Players: 2-4

Materials needed:

- Deck of cards (without face cards); combine two decks for three or four players
- Piece of paper for each player
- Pencil for each player

Object of game: To solve as many of your problems first as possible by the time all the cards are used up.

How to play

#1: Shuffle the cards. Designated dealer deals 12 cards to each player.

#2: Players keep their cards hidden from the others.

#3: All players choose 3 cards from the 12 they are holding to make a multiplication problem. The problem will be a 2-digit number times a 1-digit number. Players may use the three cards to form any numbers they like. For example, if they are holding a 2, 5, and 6 in their hand and want to use those cards to make the problem, they could decide to make 25x6, 52x6, 26x5, 56x2, 62x5, etc.

#4: Dealer asks, "Ready?" When everyone is ready, the players all set their cards down **at the same time**, making the numbers they will use for their multiplication problem. The 2-digit number will be on top, and the 1-digit number will be below it.

#5: Everyone picks up their pencils. When dealer calls, "Go!", they all work out their own problems on the paper.

#6: As soon as someone calls, "Done!", everyone else has to put their pencil down. Then they all check the player's problem and decide whether she has done it correctly.

#7: If two people call out, "Done!" at the same time, and everyone agrees that they called out at the same time, they have to check both the problems.

#8: If the other players determine that the first player to finish solved the problem correctly, the first player stacks the three cards together and sets them to the side as proof of her winning the first round.

#9: If she made an error, she can no longer participate in this turn while the other players try to finish their problems.

#10: When a player runs out of cards in his hands to use, he may draw as many as he needs from the deck to make a multiplication problem.

#11: Game ends when there are no more cards left in the stack.

How To Multiply Two 2-Digit Numbers

Do you fully understand how to multiply a two-digit number by a one-digit number? If so, you're ready to move on and get on the road to mastering multiplication both in your head and on paper.

You're going to learn how to multiply two 2-digit numbers. And here is where you are going to learn why it is so important to know how to multiply numbers that have zeros.

```
  23
 x35
```

Here's how to do it in your head. First, think of each number in its expanded form.

23 = 20 + 3
35 = 30 + 5

You are going to multiply the 20 and the 3 in 23 by the 30 and the 5 in 35. You will end up solving four multiplication problems, and adding those products together to get the final answer.

Start by multiplying 20 by 30, and then by 5.

```
  20       20
 x30      x 5
 600      100
```

Add the products together: 600 + 100 = 700.

Hold the number 700 in your mind while you finish.

Now, multiply the 3 in 23 by the 30 in 35, and then multiply the 3 by the 5 in 35.

```
   3        3
 x 30      x 5
   90       15
```

Add those products together: 90 + 15 = 105.

Now, add that sum, 105, to the number you've been holding in your head, 700:

$$105 + 700 = 805$$

There's your answer – 805. Don't believe me? Think it can't be that easy to multiply large numbers?

Okay, go get a calculator and punch in 23 X 35.

Go ahead. I'll wait.

You got 805, didn't you?

The method of multiplication I just showed you illustrates the **distributive** property.

23 X 35 = (20 + 3) X (30 + 5)

It's call the distributive property because you solve the problem by *distributing* each part of the number (digits) out to the others.

You can remember which numbers to multiply with the abbreviation FOIL: First, Outside, Inside, Last.

In other words, multiply the FIRST numbers in each addition sentence: 20 X 30.

Then, multiply the OUTSIDE numbers: 20 X 5.

Next, multiply the INSIDE numbers: 3 X 30.

Finally, multiply the LAST numbers in each addition sentence: 3 X 5.

You can use that method on paper, of course. It just is a lot to write.

Look on the next page for a way to multiply two 2-digit numbers on paper that doesn't require as much writing.

First, you are going to multiply the two digits in the units place, and write that answer below the problem.

```
  23
x 35
----
  15
```

Next, you multiply the tens place of the top number by the units place of the bottom number. Because one digit is in the tens place, you need to add a zero to the answer you get. Then, write it below the 15.

```
   23
 x 35
 ----
   15
  100
```
100 (2x5= 10; add a 0 because you're really doing 20x5)

Third, multiply the tens place in the **bottom** number by the units place in the top number. Remember to add one zero to hold the tens place. Write the number below the previous sum. In this example, that would be 100.

```
   23
 x 35
 ----
   15
  100
   90
```
90 (3x3=9, add a zero to make 90 because you are really multiplying the 30 in 35 by 3.)

Finally, multiply the tens place of the bottom number by the tens place of the top number. Since you are multiplying two tens place, you will add two zeros to your answer. Write that product below the previous sum, in this case, 90.

```
    23
  x 35
  ----
    15
   100
    90
 + 600
```
+ 600 (20X30=600)

Finally, add all those numbers that you wrote below the multiplication problem. The sum is the final product of the original numbers (in this example, 23 X 35).

15 + 100 + 90 + 600 = 805

Is there a way to solve multiplication problems with large numbers with even less writing? Sure there is! But I want you to practice this way a few times so that you will fully understand that what you are doing is solving equations using the distributive property. When you really understand the concept, then you won't make mistakes when you move to the shortcut way.

Here are eight problems for you to work on. Try to solve them in your head before you do them on paper, because the more you practice, the faster and more accurately you'll be able to do problems like this in your head.

After you solve them using the method I taught you above, check your answer on a calculator. If you got it wrong, don't beat your head against a wall! You learn by making mistakes. Just try again. Practice, not beating yourself up over mistakes, is what makes for a Math Superstar.

        42 X 76   25 X 51   36 X 18   93 X 54

        17 X 58   64 X 32   87 X 23   39 X 46

The Standard Multiplication Algorithm

Okay, do you understand how multiplying large numbers work? REALLY understand it?

Okay, I'm going to teach you the way most teachers want to see you do a multiplication problem. If you really understand the distributive property, it will make sense.

Let's stick with the original example, 23 X 35.

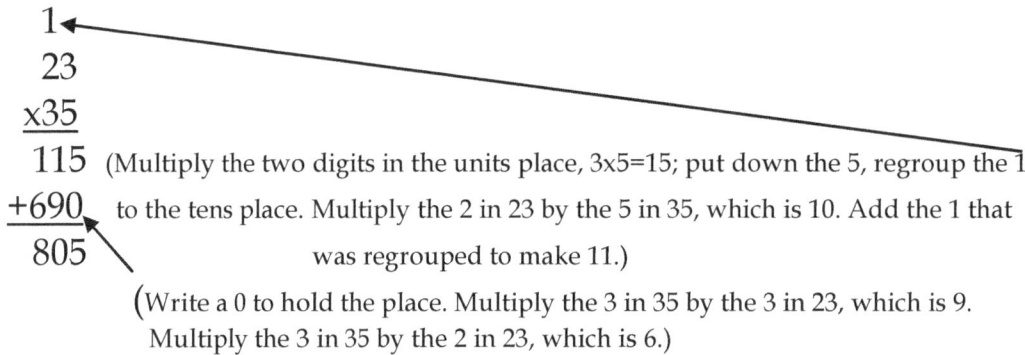

(Multiply the two digits in the units place, 3x5=15; put down the 5, regroup the 1 to the tens place. Multiply the 2 in 23 by the 5 in 35, which is 10. Add the 1 that was regrouped to make 11.)

(Write a 0 to hold the place. Multiply the 3 in 35 by the 3 in 23, which is 9. Multiply the 3 in 35 by the 2 in 23, which is 6.)

Is your brain about to explode?
MINE, TOO!

So let's both take a break. After your break, work on solving the multiplication problems I gave you on page 143 using the standard algorithm. Once you can successfully multiply two 2-digit numbers using either the long way I first showed you, or the shortcut I just taught you, then move on to the next page.

Multiplying Even LARGER Numbers

I have great news for you! Once you understand how to multiply two 2-digit numbers, you can multiply even **BIGGER** numbers – yes, even in your head (although that takes a **lot** of practice)!

The only difference is, you have more to multiply because you have more digits in at least one of the numbers.

Let's multiply a 3-digit number by a 2-digit number to get you started. We'll start by using the distributive property.

$$347$$
$$\times 53$$

$(300+40+7) \times (50+3)$

Multiply each of the numbers in the first set of parentheses by each of the numbers in the second set. FOIL still works here, but you will have an extra inside number.

Note that to do this mentally, you would want to add the first two products together and hold that sum in your head; then figure out the products of the next two problems, add them together, and add THAT number to the first sum; then figure out the products of the last two problem, add them together, then add THAT number to the sum of the other four products.

Again, we're multiplying 347x53.

$(300+40+7) \times (50+3)$

```
300x50 = 15,000
300x3  =    900
40x50  =  2,000
40x3   =    120
7x50   =    350
7x3    =     21
```

Now, add all of those answers together. You should get 18,391.

Cool, huh? Now let's do it using the long-ish paper algorithm. I'm going to start by multiplying the 3 in 53 by each of the digits in 347, keeping the place value of each digit. In other words, I am going to multiply 3 by 7, 40, and 300.

Then I am going to multiply the 50 in 53 by each of the digits in 347.

```
    347
  x  53
     21  (7x3)   ⎫
    120  (40x3)  ⎬  300, 40, and 7 in 347 all multiplied by the 3 in 53
    900  (300x3) ⎭
    350  (50x7)   ⎫
   2000  (50x40)  ⎬  300, 40, and 7 in 347 all multiplied by the 50 in 53
 +15,000 (50x300) ⎭
  18,391
```

Another way you could do it easily on paper would be to multiply 347 by 50, then 347 by 3, and add those two products together.

Following illustrates how to do that problem using the standard algorithm. Note that the 2 and the 3 in parentheses on top of the problem are the digits that are regrouped when you multiply the 5 in 53 by the digits in 347. The numbers 1 and 2 just underneath the (2) and (3) are the digits that are carried when you multiply the 3 in 53 by the digits in 347.

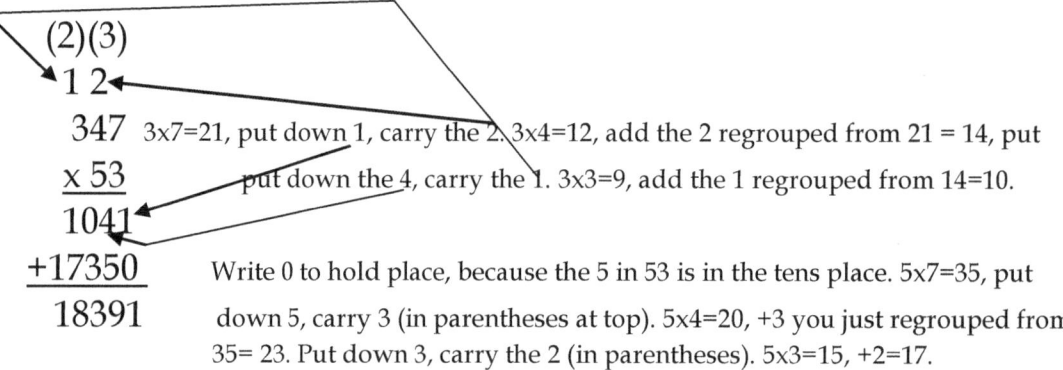

The same principles apply when you multiply two 3-digit numbers, or a 3-digit by a 4-digit number, or two 4-digit numbers. The big difference is with the standard algorithm. See how we added a zero on the line where we multiplied the 5 in 53 by each of the digits in 347? We did that because the 5 is in the tens place, so we have to stick that zero so that the first thing we multiply ends up in the tens place, too.

When you multiply a 3-digit by a 3-digit number, you will have a third line that you will start by sticking in **TWO** zeros in the units and tens place. If you ever have to multiply a 4-digit by a 4-digit number on paper, then you would have a fourth line in which you would stick **THREE** zeros.

Confused? Then, stick with using the distributive property, or long form of the algorithm (which is just a different way to use the distributive property).

If you have a really mean teacher* who is going to make you multiply two 3-digit numbers or a 3-digit by a 4-digit number on paper using the standard algorithm, you can find any number of videos on YouTube that will clarify the process. It's hard to explain it on paper!

*Okay, so maybe your teacher really isn't mean. But in real life, when grown-ups need to multiply numbers that large they either learn to do it in their heads, or they use a calculator. So why do teachers force kids to do all those tedious, brain-sucking problems on paper? Just asking. ;)

Dividing By 10, 100, And 1,000

Division is **FUN!** No, really, it is. Especially when you're dividing by ten, one hundred, or one thousand.

It's easier than slipping on a banana peel lying in a mud puddle!

Remember that division is just the opposite of multiplication. What is the rule for multiplying any number by ten?

2 X 10 = 20   54 X 10 = 540   368 X 10 = 3,680

That's right, you simply add one zero to the end of the number. In other words, you **add** an extra place value. To multiply by 100, you add two zeros, or two extra place values, and so on.

So if division is the opposite of multiplication, what do you do when you divide a number by 10, 100, or 1,000?

Not sure? Well, look at the problems below and fill in the blanks once you understand the pattern.

250 ÷ 10 = 25	4,500 ÷ 100 = 45	12,000 ÷ 1000 =
360 ÷ 10 = 36	2,700 ÷ 100 = 27	43,000 ÷ 1000 =
470 ÷ 10 = ____	1,900 ÷ 100 = ____	87,000 ÷ 1000 = ____
630 ÷ 10 = ____	5,400 ÷ 100 = ____	92,000 ÷ 1000 = ____

It's easy to divide a number by ten that ends with a zero in the units place. It's easy to divide a number by one hundred that ends with two zeros, one in the tens place and one in the units place. Or a number by 1,000 that ends with three zeros. Simply **lop off the same number of zeros that the dividend (10, 100, 1000) has,** and you have the answer.

56,000 ÷ 10 = 5,600   But what if the number you're dividing by 10, 100 or 1000 does
56,000 ÷ 100 = 560   **NOT** end in zero? What then?
56,000 ÷ 1000 = 56

Move to the next page to find out.

Division And Decimals

By now, you have probably been introduced to fractions: ¼, ½, 2/3, and so on. A fraction is a number that is greater than zero but less than one.

Another way to write a fraction is decimals. For example, the fraction 1/2 written in decimal form is 0.5. That's zero units, five **tenths.** The place to the right of the units place is called "tenths."

The place to the right of the TENTHS place, or two places to the right of a decimal place, is called "**hundredths**." So the number 0.17 is read as "seventeen hundredths." The number 0.04 is "four hundredths."

Tenths, hundredths…can you guess the name of the place to the right of the hundredths place? Units, tens, hundreds, thousands…units, tenths, hundredths…

Yes! The third place to the right of the units place and decimal is called the **thousandths** place. So the number 0.006 is read as "six thousandths". The number 0.057 is read as "fifty-seven thousandths."

Knowing all that, you can now divide **any** number by 10, 100, or 1000 – not just numbers that end with zeros. The trick is to remember that all whole numbers (any number that does not have any digits to the right of the decimal) have an invisible decimal point to the right of the units place.

When you divide by 10, 100, or 1000, you simply move the decimal point over to the right one, two, or three places and make it visible. In the following example, 453 can also be written as 453.0 . The arrows show how the decimal point is moving.

$$453 \div 10 = 45.3 \qquad 453 \div 100 = 4.53 \qquad 453 \div 1000 = .453$$

Check out the patterns below, and fill in the missing numbers once you see the pattern.

20 ÷ 10 = 2   200 ÷ 100 = 2   2000 ÷ 1000 = 2

21 ÷ 10 = 2.1   221 ÷ 100 = 2.21   2221 ÷ 1000 = 2.221

22 ÷ 10 = 2.2   222 ÷ 100 = 2.22   2222 ÷ 1000 = 2.222

23 ÷ 10 = _____   233 ÷ 100 = _____   2333 ÷ 1000 = _____

24 ÷ 10 = _____   244 ÷ 100 = _____   2444 ÷ 1000 _____

25 ÷ 10 = _____   255 ÷ 100 = _____   2555 ÷ 1000 = _____

 Now, let's try some random examples. Remember, when you divide any number by ten, simply move the decimal point over one place to the right. By 100, two places to the right. By 1000, 3 places.

I'll do the first three examples to get you started.

4 ÷ 10 = 0.4   567 ÷ 100 = 5.67   704 ÷ 1000 = 0.704

68 ÷ 10 = _____   459 ÷ 1000 = _____   751 ÷ 10 = _____

978 ÷ 100 = _____   5 ÷ 100 = _____   7,421 ÷ 1,000 = _____

47 ÷ 100 = _____   642 ÷ 10 = _____   468 ÷ 1000 = _____

Are you getting it?

YOU ROCK THE HOUSE, DUDE!

The next section on division should be a piece of cake now!

Box Division

Okay, friend, here's where the rubber meets the road. From here on out, life (that is, your life with division) will be a lot easier if you have the basic division facts memorized.

Sure, in this book, you're free to use the Multiplication Table to figure out answers. But when it comes to test time, your teacher probably won't let you use it. So work on getting those facts in your head. Go back to the pages that have the multiplication fact families and study them, say them aloud, write them down every day until they are stuck in your brain like gum in long hair.

Now that you've got those facts down, you're ready to do some serious long division, right?

What? You say that's **too hard**? Hey, are you **WHINING** at me?

I understand. **Some** people make long division out to be this **terrible monster** that only a **LEFT-BRAINED GENIUS** could ever figure out.

But what if I told you I have not one, but two methods of doing long division that won't make you want to pull your hair out? That you might even come to enjoy?

Impossible, you say? We'll see about that. Allow me to introduce Box Division.

The easiest way to teach you this method of division is to just do some examples. I'll start easy, with a 1-digit number divided by a 2-digit number.

$73 \div 5$

Since the dividend has two digits, you are going to draw a box with two columns in it. On the outside of the left of the box you write the divisor (in this case 5), and on the inside you write the dividend. The digit in the tens place, the 7, goes in the first column. The digit in the units place, the 3, goes in the second column.

5	7	3

Now you divide 5 by 7. The answer goes right above the 7.

7 ÷ 5 = 1

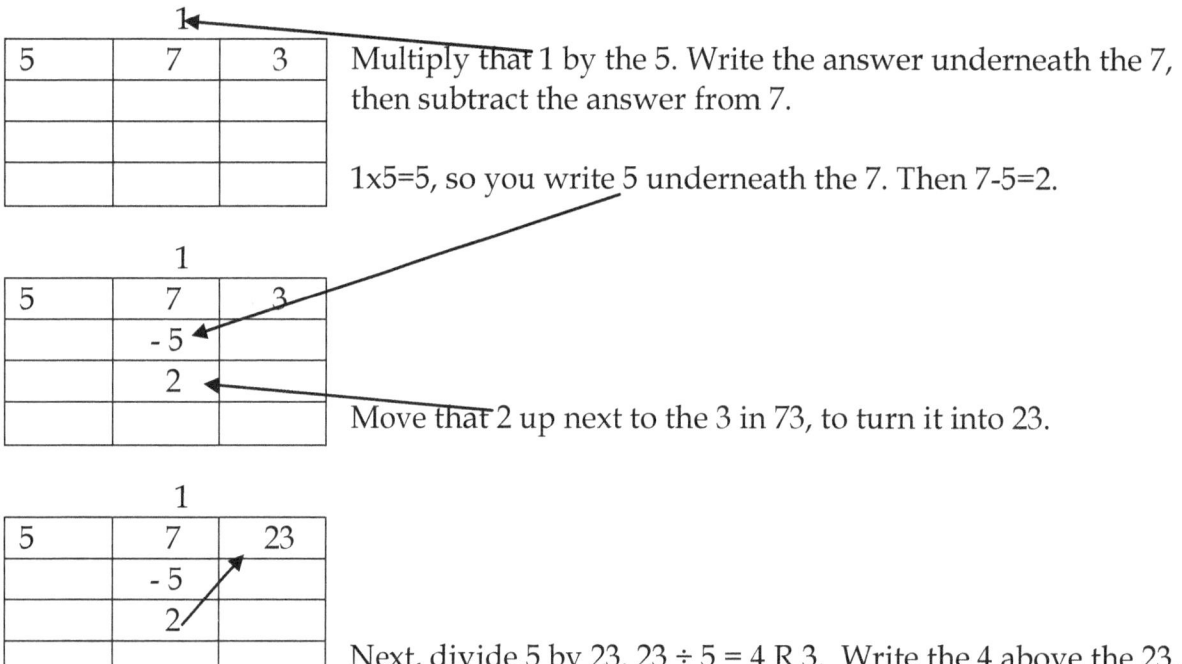

Multiply that 1 by the 5. Write the answer underneath the 7, then subtract the answer from 7.

1x5=5, so you write 5 underneath the 7. Then 7-5=2.

Move that 2 up next to the 3 in 73, to turn it into 23.

Next, divide 5 by 23. 23 ÷ 5 = 4 R 3. Write the 4 above the 23.

Then, multiply that 4 by the divisor, 5. 4x5=20. Write that 20 below the 23, and subtract it from 23.

There are no more columns of numbers to divide by 5, so you're finished! The 3 that you get when you subtract 20 from 23 is the final remainder of the answer.

73 ÷ 5 = 14 R 3

How can you be sure that's the right answer? Multiply the answer by the dividend (5), and add the remainder. It should equal the divisor (73).   14 X 5 = 70 + 3 = 73

Is that a fun way to do division, or what?

Let's try a slightly more difficult problem, a 1-digit number divided by a 3-digit number.

$$574 \div 3$$

3	5	7	4

Start with the digit in the hundreds place of the dividend, 5. Three goes into 5 only 1 time.

	1		
3	5	7	4
	-3		
	2		

1 X 3 = 3; write that 3 below the 5 and subtract.

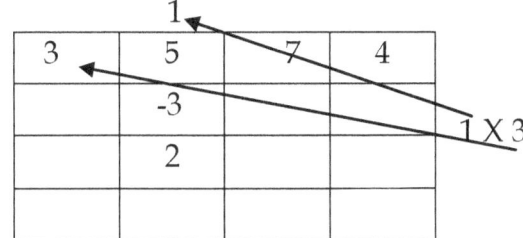

Take that 2 (from 5-3) and put it next to the 7, to turn it into 27.

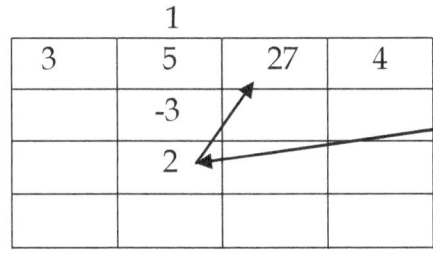

Next, figure out 27 ÷ 3. It's 9. Put that 9 above the 27.

		1	9	
3	5	27	4	
		-3	-27	
		2	0	

Multiply that 9 by the dividend, 3.   9 X 3 = 27. Write that below the 27 at the top of the box, and subtract.

		1	9	1
3	5	27	4	
		-3	-27	
		2	0	

You've divided 3 by the 5 and 7 in 574. Now, you divide the 3 by the 4.   4 ÷ 3 = 1. Write the 1 above the 4.

		1	9	1
3	5	27	4	
		-3	-27	-3
		2	0	1

Now, multiply that 1 by the 3.   1x3=3, which you write below the 4. Then, subtract the 3 from the 4.

No more columns! So we're finished, and 1 is the remainder.

574 ÷ 3 = 191 R 1. Check and make sure we did this correctly!

Does Box Division work when you're multiplying a 2-digit by a 2-digit, or a 3- or 4-digit by a 2-digit? Yes, but it ends up being as brain-crunching as using the standard division algorithm. There is a much easier way to do long division where the divisor has more than one digit.

Let's start fresh on the next page.

Division By Chunking

If you haven't understood the importance of learning to multiply by and divide by 10, you will now! The chunking method of division makes long division so easy that your parents will wonder **why on earth wasn't I taught this in school?**

The best way to develop the concept is to start simple. So let's start with 197 ÷ 3.

First, set it up like you would for the standard division algorithm.

$$3\overline{)197}$$

Then, you start **chunking** 3 into groups of 10, which is 30, and subtracting those groups from 146. You do that until you can no longer subtract 30 from the number at the bottom of the problem.

```
3)197
 - 30    10  ←3x10=30; the 10 indicates how many groups of 3 we are dividing into 197
  167
```

Can you subtract 30 from the number that is currently at the bottom of the problem, 167? Yes. So, continue on.

```
3)197
  -30    10
  167
  -30    10
  137
```

(Take a deep breath. You're doing just fine. ☺)

Can you subtract 30 from 137? Yes, so keep going. I will keep it going until the number at the bottom ends up being less than 30.

```
3)197
  -30   10
  167
  -30   10
  137
  -30   10
  107
  -30   10
   77
  -30   10
   47
  -30   10
   17
```

Well, now the number at the bottom of the problem is less than 30. So we can't divide 10 groups of 3 into it. Does that mean we're done?

Not yet! If we stopped now, that would make 17 the remainder. And what is the BIG RULE about remainders?

The remainder always has to be LESS THAN the divisor. In this example, the divisor is 3. Seventeen is greater than 3.

So, what do we do? We have to multiply 3 by a number smaller than 10, that will be equal to or less than 17.

3x6=18, so 6 is too big. But 3x5=15, which is less than 17. So we continue the problem like this:

```
3)197
  -30   10
  167
  -30   10
  137
  -30   10
  107
  -30   10
   77
  -30   10
   47
  -30   10
   17
  -15    5
    2
```

Now the number at the bottom of the problem is 2. Can we do 2 ÷ 3 and end up with a whole number for the quotient? No siree, we can't. So 2 will be the remainder for this problem.

How do we figure out the actual quotient? Add together the numbers that we wrote out to the right side of the problem:

10 + 10 + 10 + 10 + 10 + 10 + 5 = 65;   therefore,   197 ÷ 3 = 65 R 2

But don't take my word for it! Multiply 3x65, then add 2, to make sure I got the answer right.

You can use groups of 10 to chunk to solve any division problem, no matter how large. But what if you want to solve the problem more quickly? Use **multiples** of 10 – 20, 30, 40, 50, even 100 or more – to do your chunking.

Let me show you how to do that using another example. Once you understand it, you should be able to divide numbers of any digits using the chunking method, no matter how large!

Let's start with 2,345 ÷ 8.

```
8) 2,345
```

If we only used 10 groups of 8 every time, we would be subtracting 80 all the way down. And you might need several sheets of paper to do that! Because the dividend, 2345, is much greater than 80.

We could subtract 30 groups of 8, which is 240, and get the problem solved three times faster. Or we could subtract 50 groups of 8, which would be 400.

But I'm looking at the dividend and thinking that I can subtract 800 from it. So I'm going to go ahead and subtract 100 groups of 8 to start.

```
8) 2,345
   -800      100
   1545
```

1545 is greater than 800, so I'm going to chunk 100 groups of 8 again.

```
   8)2,345
     -800    100
     1545
     -800    100
      745
```

Okay, we can no longer use 100 to chunk, because 745 is less than 800. We already figured out that 50 groups of 8 is 400, so let's go ahead and use 50.

```
   8)2,345
     -800    100
     1545
     -800    100
      745
     -400     50
      345
```

Can we use 50 again? 50x8=400, and we can't subtract 400 from 345. So we have to find a smaller multiple of ten that will work.

How about 40? 40x8=320. Perfect! That's very close to the number we are subtracting from, 345, but less than it.

```
   8)2,345
     -800    100
     1545
     -800    100
      745
     -400     50
      345
     -320     40
       25
```

How many groups of 8 go into 25? 3, because 8x3=24.

```
8)2,345
  -800    100
  1545
  -800    100
   745
  -400     50
   345
  -320     40
    25
   -24      3
     1
```

How many times does 8 go into 1? Can you do it and end up with a whole number as an answer?

Nope. 1 is less than 8, so we're done, and 1 is the remainder.

So what will be the quotient? Do you remember how to figure out the quotient with the chunking method?

Yes, you add together the numbers at the right side of the problem, all the numbers that you multiplied by 8.

100 + 100 + 50 + 40 + 3 = 293

2,345 ÷ 8 = 293 R 1 (the remainder is the number left at the very bottom of the problem, if it is greater than 0)

Isn't that the

Let me show you how to do a similar problem using chunking that involves a 2-digit divisor.

$$42\overline{)7{,}201}$$

Again, using 100 to chunk will work at the very beginning. Then I will use smaller multiples of ten to chunk.

```
42)7,201
  -4,200   100
   3,001
  -2,100    50   I chose 50 because I knew that 100x42=4200, and that half of 4200=2100.
     901         Half of 100 is 50, so I knew that 50 groups of 42 would be 2100, which is
    -420    10                        close to, but still less than, 3,001.
     481
    -420    10
      61
     -42     1
      19
```

19 is less than 42, so we're finished, and 19 will be the remainder. The quotient will be:

100 + 50 + 10 + 10 + 1 = 171

7,201 ÷ 42 = 171 R 19

You have now learned long division, and can graduate from high school!

Okay, maybe not quite yet. But at least we're **ALL DONE** with long division, right?

I'd really like to be. But I need to cover all my bases.

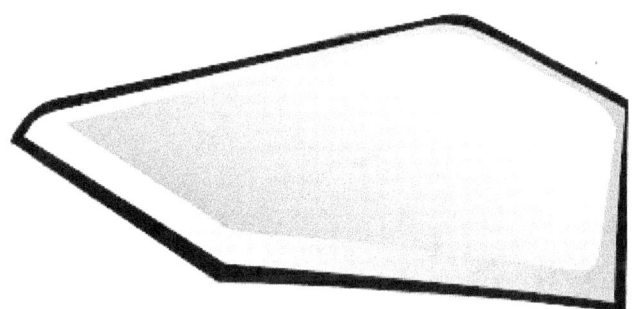

(That means, I need to make happy all the grown-ups in your life who are in charge of your education.) So I need to explain to you the standard division algorithm.

I know, I know, boring, snoring. But you have to admit that up to this point the book has been pretty *FUN*, right?

But before I move on, I want to let you know if my paper explanations for either the box division method or the chunking method were confusing, you can find videos on YouTube that demonstrate those methods. They might be easier to follow. Just do a search.

Okay, now on to the **least fun** part of this book (but we're **ALMOST DONE**! ☺ ☺ ☺)

The Standard Division Algorithm

What is 73 ÷ 9?

That one is easy. You know that 9 X 8 = 72. So 73 ÷ 9 = 8 R 1. We've already practiced a bunch of problems like that.

But what about 73 ÷ 6? Here's how you lay it out.

$$6\overline{)73}$$

After you set up the problem like that, you begin a game of 20 questions with yourself.

Question 1: Can I divide 6 into 7 end up with a number equal to or greater than 1 (a whole number) for the answer?

Answer: Yes! If the divisor (7) is greater than the dividend (6), I can divide it and end up with a whole number (1 or a bigger number).

Question 2: So, what is 7 ÷ 6?

Answer: 1, with a remainder of 1. So, write a 1 above the 7 in the problem.

$$\begin{array}{r} 1 \\ 6\overline{)73} \end{array}$$

Next, you multiply that 1 times the divisor, 6, and write the answer **below** the seven. (You are actually multiplying 10 X 6, since the 1 is in the tens place).

$$\begin{array}{r} 1 \\ 6\overline{)73} \\ 6 \end{array}$$

After that, you subtract the 6 (60, really) from the 7 (70).

$$\begin{array}{r} 1 \\ 6\overline{)73} \\ \underline{-6} \\ 1 \end{array}$$

Question 3: Can I divide 6 into 1 and end up with a whole number as the answer?

Answer: No. 6 ÷ 1 will equal a decimal, or a number **less** than 1.

So, you have to bring down the 3 in 73 to make a number larger than 6.

$$\begin{array}{r} 1 \\ 6\overline{)73} \\ -6\downarrow \\ \hline 13 \end{array}$$

Question 4: Can I divide 6 into 13 and end up with a whole number?

Answer: Yes. 13 ÷ 6 = 2 (R 1) [6x2=12, add 1 to get 13.]

Now write the 2 above the 3 in 73.

$$\begin{array}{r} 12 \\ 6\overline{)73} \\ -6 \\ \hline 13 \end{array}$$

Multiply that 2 times the divisor, 6, and write the answer (12) below the 13.

$$\begin{array}{r} 12 \\ 6\overline{)73} \\ -6 \\ \hline 13 \\ -12 \end{array}$$

Subtract the 12 from the 13.

$$\begin{array}{r} 12 \\ 6\overline{)73} \\ -6 \\ \hline 13 \\ -12 \\ \hline 1 \end{array}$$

Question 5: Can I divide 6 into that 1 at the bottom of the problem?

Answer: No, I can't.

Question 6: Are there any more digits I can bring down to turn it into a larger number that I __can__ divide by 6?

Answer: No. That means, I'm almost done with the problem.

```
      12 R 1
   6 | 73
      -6
      13
     -12
       1
```

After the 12, I write "R 1" because there is a remainder of 1 that I cannot divide six into. Remember: **the remainder must always be smaller than the divisor in the original problem.**

$$73 \div 6 = 12 \text{ R } 1$$

Let's do one more 2-digit divided by 1-digit problem using the standard algorithm.

```
   4 | 87
```

Can I divide 4 into 8? $8 \div 4 = ??$

Yes. $8 \div 4 = 2$

```
       2
   4 | 87
```

$2 \times 4 = 8$, so I write an 8 underneath the 8 in 87 (I'm really multiplying 20x4 to get 80).

```
       2
   4 | 87
      -8
```

Subtract 8 from 8.

```
       2
   4 | 87
      -8
       0
```

I can't divide 4 into 0, so I have to bring down the 7 in 87.

```
   2
4)87↓
  -8↓
   07
```

Can I do 7 ÷ 4 (or, divide 4 into 1)? Yes. It will be 1 with a remainder of 3, because 4x1=4, plus 3 equals 7. So I write the 1 above the 7 in 87.

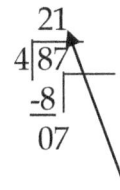

Now I multiply that 1 times the divisor, which is 4, and write the answer underneath the 7.

```
   21
4)87
  -8
   07
   -4
```

Next, I subtract 4 from 7.

```
   21
4)87
  -8
   07
   -4
    3
```

Can I divide 4 into 3 (3 ÷ 4) and end up with a whole number? No, I can't. Can I bring any other digits down from the dividend (87)? Nope.

So, I'm done except for one more thing.

```
       21 R 3   → Here is the answer to 87 ÷ 4.
    4)87
     -8
      07
      -4
       3
```

Below are a few problems for you to practice this skill. Remember that if you end up with a remainder, *it must always be smaller than the original divisor.*

Also, to check your answers to make sure they are correct, change the division problem into a multiplication problem.

```
   21
   X 4
   84, 84+3 (the remainder)=87
```

Here are some problems for you to try on a separate sheet of paper.

23 ÷ 3   45 ÷ 2   89 ÷ 7   52 ÷ 4

95 ÷ 9   39 ÷ 2   98 ÷ 7   53 ÷ 8

Once you understand how to do 2-digit divided by 1-digit number problems, you can do any long division problem where the divisor is only 1-digit.

The tricky part with the standard algorithm for long division comes when you have to divide two-digit numbers by three- or four-digit numbers. The process is the same as for the above problems, but it takes a little more multiplication know-how and a little more brain power.

To help you get started with that, on the next page I will demonstrate a two-digit divided by four-digit problem.

$$62 \overline{\smash{)}4{,}927}$$

I'm dealing with a 2-digit divisor, 62. So I'm going to start out by asking if I can divide 62 into the first two digits of the dividend, 49.

No, 62 cannot divide into 49 and equal a number equal to or greater than 1. So I ask myself if I can divide 62 into 492.

Yes, because 492 is obviously greater than 62. How much greater?

It's less than 10 times greater, because 62 X 10 = 620. How about 5 times greater? Because 5 times greater would be half of 620, or 310. Let's try 5.

The 5 will go above the digit 2 in 4,927 because we are dividing 62 in 492.

$$62 \overline{\smash{)}4{,}927}^{5}$$

Next, I multiply 62 X 5, then write the product below 492 and subtract it from 492.

$$\begin{array}{r} 5 \\ 62 \overline{\smash{)}4{,}927} \\ 310 \end{array}$$

OOPS! I can tell already that's not going to work. How? 492 is very close to 500, and 310 is about 300. 500-300=200, which is way bigger than 62. The number I subtract from 492 needs to give me a number that is less than 62.

Hmm. Let me think about it. I know! Let me just look at the 6 in 62, and the 49 in 4,927. What can I multiply by 6 to get close to 49? Well, 7x6=42, and 8x6=48. I'm going to try 8, because 62 is very close to 60 so I don't think 62 X 8 will go over 492. Maybe. Let me try.

$$\begin{array}{r} 8 \\ 62 \overline{\smash{)}4{,}927} \\ 496 \end{array}$$

Doggone it! Eight is just a little bit too big, because I can't subtract 496 from 492. Seven it will be.

$$\begin{array}{r} 7 \\ 62\overline{)4927} \\ 434 \end{array}$$

Yay! I did it! Now I can subtract…

$$\begin{array}{r} 7 \\ 62\overline{)4927} \\ \underline{-434} \\ 58 \end{array}$$

YES! 58 is less than 62, so I got it right. But now I can't divide any further, because 58 is less than 62.

Hold on. I've still got that 7 in the units place of the dividend that I can bring down. So I need to do that, and the figure out how many times 62 can go into the resulting number.

$$\begin{array}{r} 7 \\ 62\overline{)4927} \\ \underline{-434} \\ 587 \end{array}$$

So my next thought will be, how many times can 6 go into 58? Well, 9x6=54, so let me try 9.

$$\begin{array}{r} 79 \\ 62\overline{)4927} \\ \underline{-434} \\ 587 \\ 558 \end{array}$$

Yes! That worked! 9x62=558, which is less than 587, so I can subtract it from 587. The only thing is, will I end up with a number less than 62?

$$
\begin{array}{r}
79 \\
62\overline{\smash{)}4927} \\
-434 \\
\hline
587 \\
-558 \\
\hline
29
\end{array}
$$

AWESOME! I did it! I ended up with a number less than 62. I don't have any more digits to bring down, so all I have to do now is write the remainder next to the quotient, 79.

$$
\begin{array}{r}
79 \text{ R } 29 \\
62\overline{\smash{)}4927}\phantom{\text{ R 29}} \\
-434\phantom{0\text{ R 29}} \\
\hline
587\phantom{\text{ R 29}} \\
-558\phantom{\text{ R 29}} \\
\hline
29\phantom{\text{ R 29}}
\end{array}
$$

And there you go! How to divide using the standard algorithm (but can you see why I prefer the chunking method?).

You finished this book! You are now officially a

Now, go celebrate! Tell your best friend! Ask your mom to bake you some brownies! Take the rest of the day off to play with your dog or at the park or to read your favorite book!

I'm **SO** proud of you…and you should be incredibly, extremely, marvelously, fantastically pleased with yourself! ☺ ☺ ☺ ☺

****

PARENTS/TEACHER:

Please remember to help me out with a review if you have enjoyed this book.

Thank you! ☺

Emily Jacques

www.ingramcontent.com/pod-product-compliance
Lightning Source LLC
Chambersburg PA
CBHW080009210526
45170CB00015B/1953